WIRING
ESSENTIALS

BLACK&
DECKER®
QUICK
STEPS™

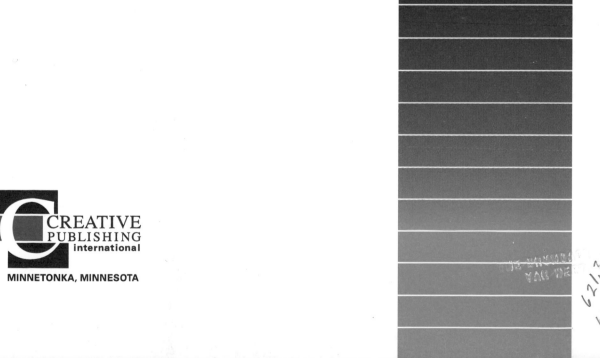

CREATIVE
PUBLISHING
international

MINNETONKA, MINNESOTA

Credits

Copyright © 1996
Creative Publishing International, Inc.
5900 Green Oak Drive
Minnetonka, Minnesota 55343
1-800-328-3895
All rights reserved
Printed in U.S.A.

CREATIVE PUBLISHING international

President: Iain Macfarlane

Created by: The Editors of Creative Publishing International, Inc., in cooperation with Black & Decker.  is a trademark of the Black & Decker Corporation and is used under license.

Printed on American paper by:
Quebecor Printing
02 01 00 99 98 / 5 4 3 2

Books available in this series:

Wiring Essentials
Plumbing Essentials
Carpentry Essentials
Painting Essentials
Flooring Essentials
Landscape Essentials
Masonry Essentials
Door & Window Essentials
Roof System Essentials
Deck Essentials
Porch & Patio Essentials
Built-In Essentials

Contents

Faucet

Water flows
under pressure

Water supply pipe

Drain pipe

Water returns
under no pressure

Understanding Electricity

A household electrical system can be compared with a home's plumbing system. Electrical current flows in wires in much the same way that water flows inside pipes. Both electricity and water enter the home, are distributed throughout the house, do their "work," and then exit.

In plumbing, water first flows through the pressurized water supply system. In electricity, current first flows along hot wires. Current flowing along hot wires also is pressurized. The pressure of electrical current is called **voltage**.

Large supply pipes can carry a greater volume of water than small pipes. Likewise, large electrical wires carry more current than small wires. This current-carrying capacity of wires is called **amperage**.

Water is made available for use through the faucets, spigots, and shower heads in a home. Electricity is made available through receptacles, switches, and fixtures.

Water finally leaves the home through a drain system, which is not pressurized. Similarly, electrical current flows back through neutral wires. The current in neutral wires is not pressurized, and is said to be at zero voltage.

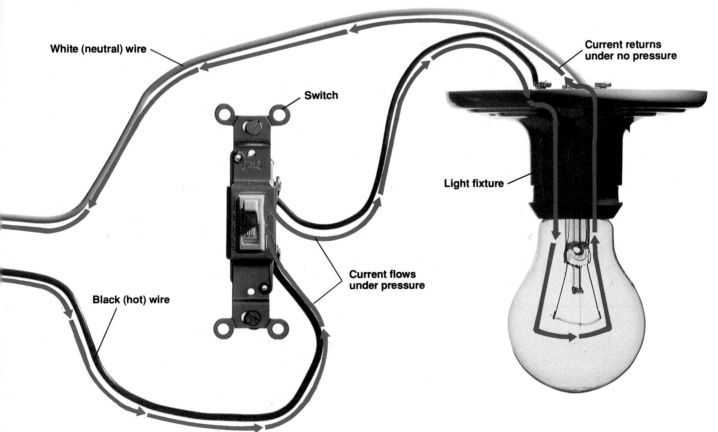

White (neutral) wire

Current returns
under no pressure

Switch

Light fixture

Current flows
under pressure

Black (hot) wire

Glossary of Electrical Terms

ampere (or **amp**): Refers to the rate at which electrical power flows to a light, tool, or an appliance.

armored cable: Two or more wires that are grouped together and protected by a flexible metal covering.

box: A device used to contain wiring connections.

BX: See **armored cable.**

cable: Two or more wires that are grouped together and protected by a covering or sheath.

circuit: A continuous loop of electrical current flowing along wires or cables.

circuit breaker: A safety device that interrupts an electrical circuit in the event of an overload or short circuit.

conductor: Any material that allows electrical current to flow through it. Copper wire is an especially good conductor.

conduit: A metal or plastic tube used to protect wires.

continuity: An uninterrupted electrical pathway through a circuit or electrical fixture.

current: The movement of electrons along a conductor.

duplex receptacle: A receptacle that provides connections for two plugs.

feed wire: A conductor that carries 120-volt current uninterrupted from the service panel.

fuse: A safety device, usually found in older homes, that interrupts electrical circuits during an overload or short circuit.

Greenfield: See **armored cable.**

grounded wire: See **neutral wire.**

grounding wire: A wire used in an electrical circuit to conduct current to the earth in the event of a short circuit. The grounding wire often is a bare copper wire.

hot wire: Any wire that carries voltage. In an electrical circuit, the hot wire usually is covered with black or red insulation.

insulator: Any material, such as plastic or rubber, that resists the flow of electrical current. Insulating materials protect wires and cables.

junction box: See **box.**

meter: A device used to measure the amount of electrical power being used.

neutral wire: A wire that returns current at zero voltage to the source of electrical power. Usually covered with white or light gray insulation. Also called the grounded wire.

outlet: See **receptacle.**

overload: A demand for more current than the circuit wires or electrical device was designed to carry. Usually causes a fuse to blow or a circuit breaker to trip.

pigtail: A short wire used to connect two or more circuit wires to a single screw terminal.

polarized receptacle: A receptacle designed to keep hot current flowing along black or red wires, and neutral current flowing along white or gray wires.

power: The result of hot current flowing for a period of time. Use of power makes heat, motion, or light.

receptacle: A device that provides plug-in access to electrical power.

Romex: A brand name of plastic-sheathed electrical cable that is commonly used for indoor wiring.

screw terminal: A place where a wire connects to a receptacle, switch, or fixture.

service panel: A metal box usually near the site where electrical power enters the house. In the service panel, electrical current is split into individual circuits. The service panel has circuit breakers or fuses to protect each circuit.

short circuit: An accidental and improper contact between two current-carrying wires, or between a current-carrying wire and a grounding conductor.

switch: A device that controls electrical current passing through hot circuit wires. Used to turn lights and appliances on and off.

UL: An abbreviation for Underwriters Laboratories, an organization that tests electrical devices and manufactured products for safety.

voltage (or **volts**): A measurement of electricity in terms of pressure.

wattage (or **watt**): A measurement of electrical power in terms of total energy consumed. Watts can be calculated by multiplying the voltage times the amps.

wire nut: A device used to connect two or more wires together.

Electricity & Safety

Safety should be the primary concern of anyone working with electricity. Although most household electrical repairs are simple and straightforward, always use caution and good judgment when working with electrical wiring or devices. Common sense can prevent accidents.

The basic rule of electrical safety is: **Always turn off power to the area or device you are working on.** At the main service panel, remove the fuse or shut off the circuit breaker that controls the circuit you are servicing. Then check to make sure

the power is off by testing for power with a neon circuit tester (page 14). Restore power only when the repair or replacement project is complete.

Follow the safety tips shown on these pages. Never attempt an electrical project beyond your skill or confidence level. Never attempt to repair or replace your main service panel or service entrance head. These are jobs for a qualified electrician, and require that the power company shuts off power to your house.

Shut off power to the proper circuit at the fuse box or main service panel before beginning work.

Make a map of your household electrical circuits to help you turn the proper circuits on and off for electrical repairs.

Close service panel door and post a warning sign to prevent others from turning on power while you are working on electrical projects.

Keep a flashlight near your main service panel. Check flashlight batteries regularly.

Always check for power at the fixture you are servicing before you begin any work.

Use only UL approved electrical parts or devices. These devices have been tested for safety by Underwriters Laboratories.

Wear rubber-soled shoes while working on electrical projects. On damp floors, stand on a rubber mat or dry wooden boards.

Use fiberglass or wood ladders when making routine household repairs near the service head.

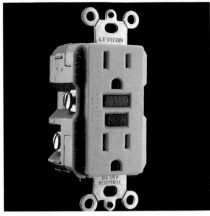

Use GFCI receptacles (ground-fault circuit-interrupters) where specified by local electrical codes (pages 62 to 63).

Protect children with receptacle caps or childproof receptacle covers.

Use extension cords only for temporary connections. Never place them underneath rugs, or fasten them to walls, baseboards, or other surfaces.

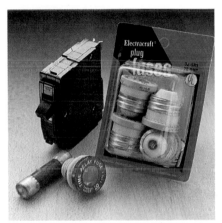

Use correct fuses or breakers in the main service panel (pages 24 to 25). Never install a fuse or breaker that has a higher amperage rating than the circuit wires.

Do not touch metal pipes, faucets, or fixtures while working with electricity. The metal may provide a grounding path, allowing electrical current to flow through your body.

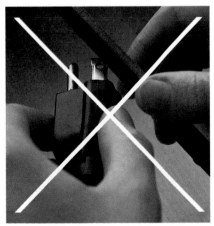

Never alter the prongs of a plug to fit a receptacle. If possible, install a new grounded receptacle.

Do not drill walls or ceilings without first shutting off electrical power to the circuits that may be hidden. Use double-insulated tools.

Your Electrical System

Electrical power that enters the home is produced by large **power plants**. Power plants are located in all parts of the country, and generate electricity with turbines that are turned by water, wind, or steam. From these plants electricity enters large "step-up" transformers that increase voltage to half a million volts or more.

Electricity flows easily at these large voltages, and travels through high-voltage transmission lines to communities that can be hundreds of miles from the power plants. "Step-down" transformers located at **substations** then reduce the voltage for distribution along street lines. On **utility power poles**, smaller transformers further reduce the voltage to ordinary 120-volt current for household use.

Lines carrying current to the house either run underground, or are strung overhead and attached to a post called a **service head**. Most homes built after 1950 have three wires running to the service head: two power lines, each carrying 120 volts of current, and a grounded neutral wire. Power from the two 120-volt lines may be combined at the **service panel** to supply current to large, 240-volt appliances like clothes dryers or electric water heaters.

Many older homes have only two wires running to the service head, with one 120-volt line and a grounded neutral wire. This older two-wire service is inadequate for today's homes. Contact an electrical contractor and your local power utility company to upgrade to a three-wire service.

Incoming power passes through an **electric meter** that measures power consumption. Power then enters the service panel, where it is distributed to circuits that run throughout the house. The service panel also contains fuses or circuit breakers that shut off power to the individual circuits in the event of a short circuit or an overload. Certain high-wattage appliances, like microwave ovens, are usually plugged into their own individual circuits to prevent overloads.

Voltage ratings determined by power companies and manufacturers have changed over the years. Current rated at 110 volts changed to 115 volts, then 120 volts. Current rated at 220 volts changed to 230 volts, then 240 volts. Similarly, ratings for receptacles, tools, light fixtures, and appliances have changed from 115 to 125 volts. These changes will not affect the performance of new devices connected to older wiring. For making electrical calculations, such as the ones shown in "Evaluating Circuits for Safe Capacity" (pages 26 to 27), use a rating of 120 volts or 240 volts for your circuits.

Power plants supply electricity to thousands of homes and businesses. Step-up transformers increase the voltage produced at the plant, making the power flow more easily along high-voltage transmission lines.

Substations are located near the communities they serve. A typical substation takes current from high-voltage transmission lines and reduces it for distribution along street lines.

Utility pole transformers reduce the high-voltage current that flows through power lines along neighborhood streets. A utility pole transformer reduces voltage from 10,000 volts to the normal 120-volt current used in households.

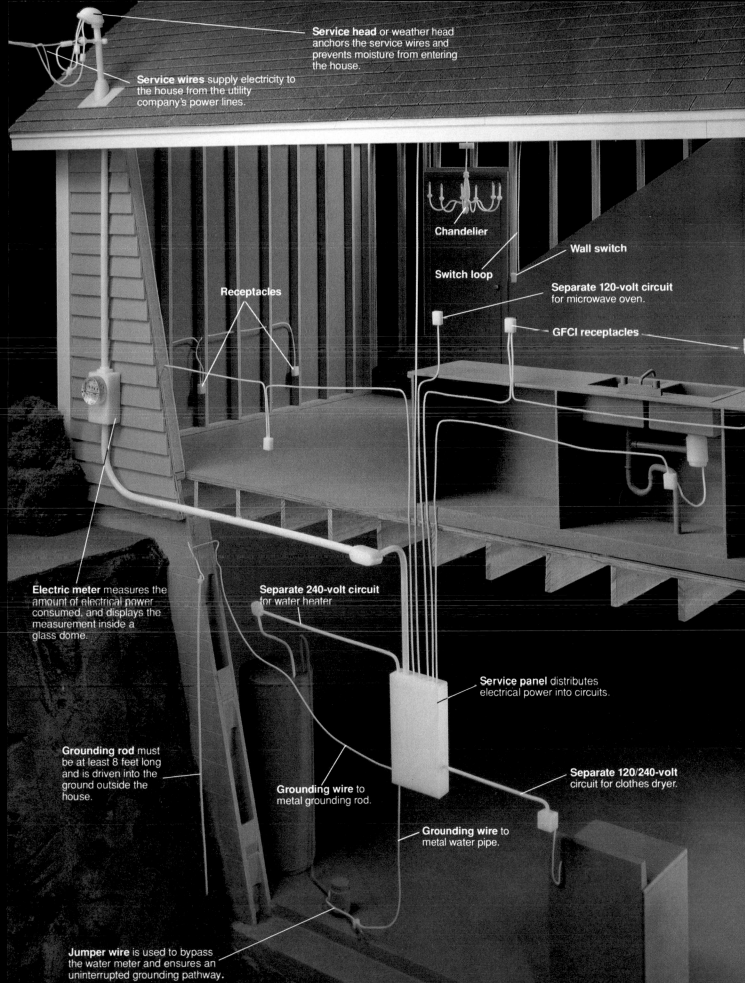

Service head or weather head anchors the service wires and prevents moisture from entering the house.

Service wires supply electricity to the house from the utility company's power lines.

Chandelier

Switch loop

Wall switch

Separate 120-volt circuit for microwave oven.

GFCI receptacles

Receptacles

Electric meter measures the amount of electrical power consumed, and displays the measurement inside a glass dome.

Separate 240-volt circuit for water heater

Service panel distributes electrical power into circuits.

Grounding rod must be at least 8 feet long and is driven into the ground outside the house.

Grounding wire to metal grounding rod.

Separate 120/240-volt circuit for clothes dryer.

Grounding wire to metal water pipe.

Jumper wire is used to bypass the water meter and ensures an uninterrupted grounding pathway.

Understanding Circuits

An electrical circuit is a continuous loop. Household circuits carry power from the main service panel, throughout the house, and back to the main service panel. Several switches, receptacles, light fixtures, or appliances may be connected to a single circuit.

Current enters a circuit loop on hot wires, and returns along neutral wires. These wires are color coded for easy identification. Hot wires are black or red, and neutral wires are white or light gray. For safety, most circuits include a bare copper or green insulated grounding wire. The grounding wire conducts current in the event of a short circuit or overload, and helps reduce the chance of severe electrical shock. The service panel also has a grounding wire connected to a metal water pipe and metal grounding rod buried underground (pages 12 to 13).

If a circuit carries too much power, it can overload. A fuse or circuit breaker protects each circuit in case of overloads (pages 24 to 25). To calculate how much power any circuit can carry, see "Evaluating Circuits for Safe Capacity" (pages 26 to 27).

Current returns to the service panel along a neutral circuit wire. Current then becomes part of a main circuit and leaves the house on a large neutral service wire that returns it to the utility pole transformer.

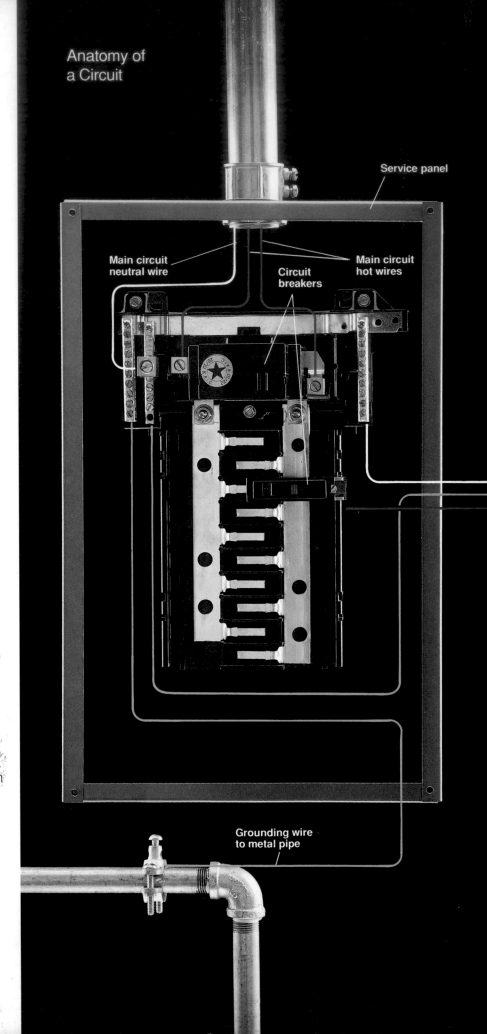

Anatomy of a Circuit

Service panel

Main circuit neutral wire

Circuit breakers

Main circuit hot wires

Grounding wire to metal pipe

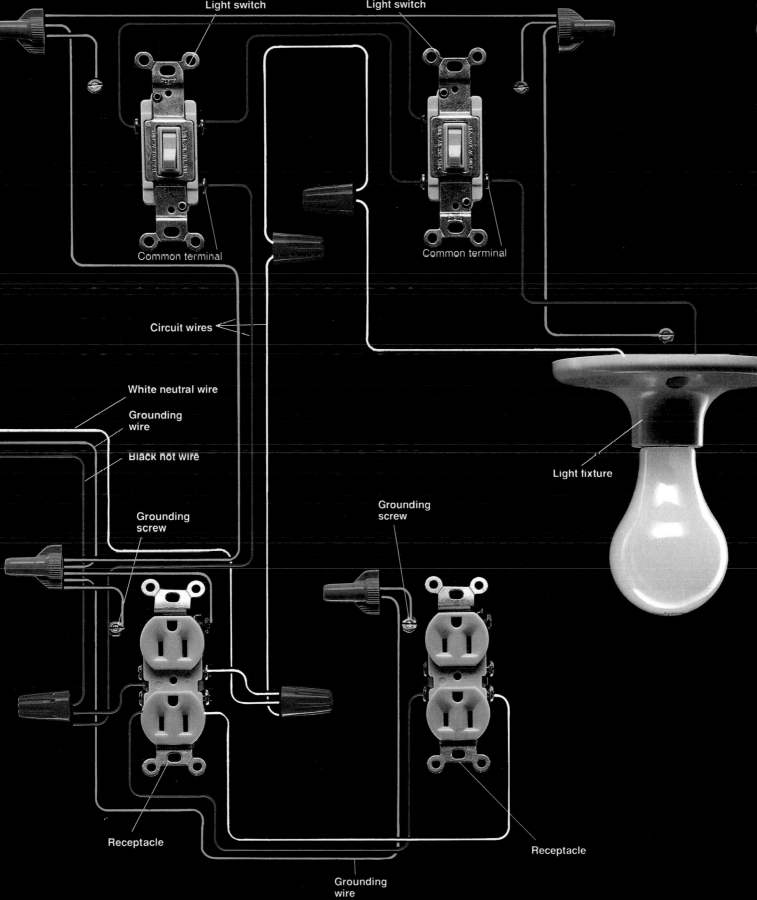

Light switch

Light switch

Common terminal

Common terminal

Circuit wires

White neutral wire

Grounding wire

Black hot wire

Light fixture

Grounding screw

Grounding screw

Receptacle

Receptacle

Grounding wire

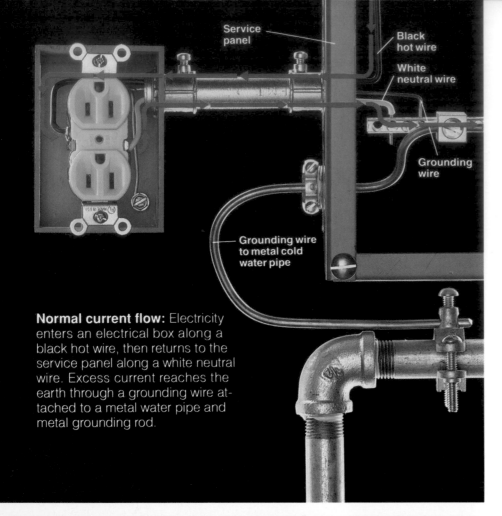

Service panel

Black hot wire

White neutral wire

Grounding wire

Grounding wire to metal cold water pipe

Normal current flow: Electricity enters an electrical box along a black hot wire, then returns to the service panel along a white neutral wire. Excess current reaches the earth through a grounding wire attached to a metal water pipe and metal grounding rod.

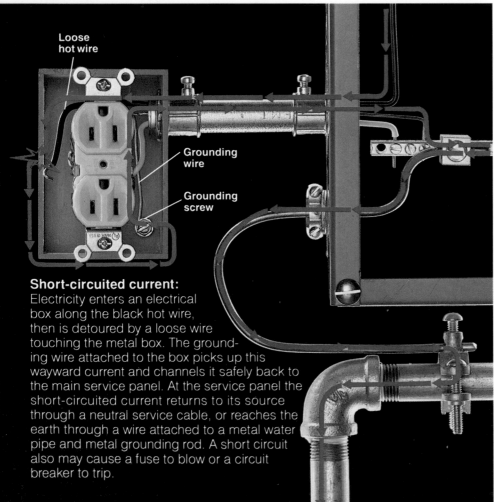

Loose hot wire

Grounding wire

Grounding screw

Short-circuited current:
Electricity enters an electrical box along the black hot wire, then is detoured by a loose wire touching the metal box. The grounding wire attached to the box picks up this wayward current and channels it safely back to the main service panel. At the service panel the short-circuited current returns to its source through a neutral service cable, or reaches the earth through a wire attached to a metal water pipe and metal grounding rod. A short circuit also may cause a fuse to blow or a circuit breaker to trip.

Grounding & Polarization

Electricity always seeks to return to its source and complete a continuous circuit. In a household wiring system, this return path is provided by white neutral wires that return current to the main service panel. From the service panel, current returns along a neutral service wire to a power pole transformer.

A **grounding wire** provides an additional return path for electrical current. The grounding wire is a safety feature. It is designed to conduct electricity if current seeks to return to the service panel along a path other than the neutral wire, a condition known as a **short circuit**.

A short circuit is a potentially dangerous situation. If an electrical box, tool or appliance becomes short circuited and is touched by a person, the electrical current may attempt to return to its source by passing through that person's body.

However, electrical current always seeks to move along the easiest path. A grounding wire provides a safe, easy path for current to follow back to its source. If a person touches an electrical box, tool, or appliance that has a properly installed grounding wire, any chance of receiving a severe electrical shock is greatly reduced.

In addition, household wiring systems are required to be connected directly to the earth. The earth has a unique ability to absorb the electrons of electrical current. In the event of a short circuit or overload, any excess electricity will find its way along the grounding wire to the earth, where it becomes harmless.

This additional grounding is completed by wiring the household electrical system to a metal cold water pipe and a metal grounding rod that is buried underground.

After 1920, most American homes included receptacles that accepted **polarized plugs**. While not a true grounding method, the two-slot polarized plug and receptacle was designed to keep hot current flowing along black or red wires, and neutral current flowing along white or gray wires.

Armored cable and metal conduit, widely installed in homes during the 1940s, provided a true grounding path. When connected to metal junction boxes, it provided a metal pathway back to the service panel.

Modern cable includes a bare or green insulated copper wire that serves as the grounding path. This grounding wire is connected to all receptacles and metal boxes to provide a continuous pathway for any short-circuited current. A cable with a grounding wire usually is attached to three-slot receptacles. By plugging a three-prong plug into a grounded three-slot receptacle, appliances and tools are protected from short circuits.

Use a receptacle adapter to plug three-prong plugs into two-slot receptacles, but use it only if the receptacle connects to a grounding wire or grounded electrical box. Adapters have short grounding wires or wire loops that attach to the receptacle's coverplate mounting screw. The mounting screw connects the adapter to the grounded metal electrical box.

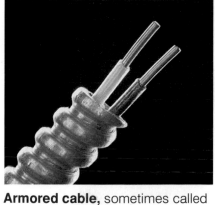

Modern NM (nonmetallic) cable, found in most wiring systems installed after 1965, contains a bare copper wire that provides grounding for receptacle and switch boxes.

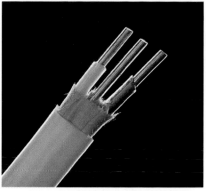

Armored cable, sometimes called BX or Greenfield cable, has a metal sheath that serves as the grounding pathway. Short-circuited current flows through the metal sheath back to the service panel.

Polarized receptacles have a long slot and a short slot. Used with a polarized plug, the polarized receptacle keeps electrical current directed for safety.

Three-slot receptacles are required by code for new homes. They are usually connected to a standard two-wire cable with ground (above, left).

Receptacle adapter allows three-prong plugs to be inserted into two-slot receptacles. The adapter can be used only with grounded receptacles, and the grounding loop or wire of the adapter must be attached to the coverplate mounting screw of the receptacle.

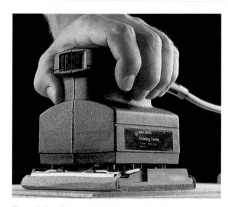

Double-insulated tools have nonconductive plastic bodies to prevent shocks caused by short circuits. Because of these features, double-insulated tools can be used safely with ungrounded receptacles.

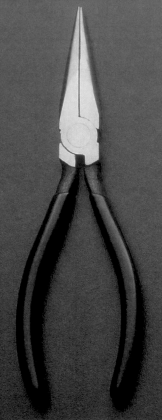

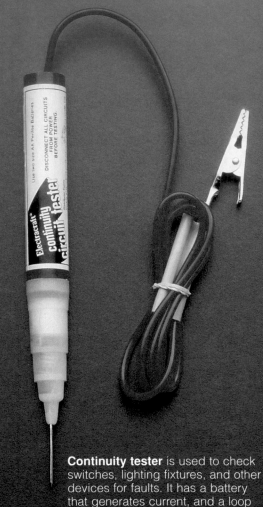

Needlenose pliers bends and shapes wires for making screw terminal connections. Some needlenose pliers also have cutting jaws for clipping wires.

Combination tool is essential for home wiring projects. It cuts cables and individual wires, measures wire gauges, and strips the insulation from wires. It has insulated handles.

Continuity tester is used to check switches, lighting fixtures, and other devices for faults. It has a battery that generates current, and a loop of wire for creating an electrical circuit.

Cordless screwdriver drives a wide variety of screws and fasteners. It is rechargeable, and can be used in either a power or manual mode. A removable tip allows the cordless screwdriver to drive either slotted or Phillips screws.

Neon circuit tester is used to check circuits wires for power. Testing for power is an essential safety step in any electrical repair project.

Tools for Electrical Repairs

Home electrical repairs require only a few inexpensive tools. As with any tool purchase, invest in quality when buying tools for electrical repairs.

Keep tools clean and dry, and store them securely. Tools with cutting jaws, like needlenose pliers and combination tools, should be resharpened or discarded if the cutting edges become dull.

Several testing tools are used in electrical repair projects. Neon circuit testers (page 58), continuity testers (page 42), and multi-testers (below) should be checked periodically to make sure they are operating properly. Continuity testers and multi-testers have batteries that should be replaced regularly.

Insulated screwdrivers have rubber-coated handles that reduce the risk of shock if the screwdriver should accidentally touch live wires.

Fuse puller is used to remove cartridge-type fuses from the fuse blocks usually found in older main service panels.

Cable ripper fits over NM (nonmetallic) cable. A small cutting point rips the outer plastic vinyl sheath on NM cable so the sheath can be removed without damaging wires.

Multi-tester is a versatile, battery-operated tool frequently used to measure electrical voltages. It also is used to test for continuity in switches, light fixtures, and other electrical devices. An adjustable control makes it possible to measure current ranging from 1 to 1000 volts. A multi-tester is an essential tool for measuring current in low-voltage transformers, like those used to power doorbell and thermostat systems.

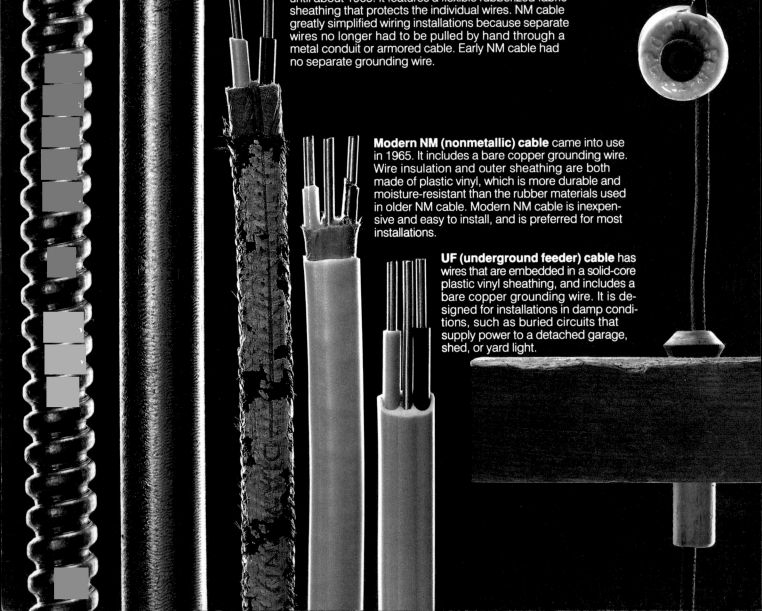

until about 1965. It features a flexible rubberized fabric sheathing that protects the individual wires. NM cable greatly simplified wiring installations because separate wires no longer had to be pulled by hand through a metal conduit or armored cable. Early NM cable had no separate grounding wire.

Modern NM (nonmetallic) cable came into use in 1965. It includes a bare copper grounding wire. Wire insulation and outer sheathing are both made of plastic vinyl, which is more durable and moisture-resistant than the rubber materials used in older NM cable. Modern NM cable is inexpensive and easy to install, and is preferred for most installations.

UF (underground feeder) cable has wires that are embedded in a solid-core plastic vinyl sheathing, and includes a bare copper grounding wire. It is designed for installations in damp conditions, such as buried circuits that supply power to a detached garage, shed, or yard light.

Wires & Cables

Wires are made of copper, aluminum, or aluminum covered with a thin layer of copper. Solid copper wires are the best conductors of electricity and are the most widely used. Aluminum and copper-covered aluminum wires require special installation techniques. They are discussed on page 18.

A group of two or more wires enclosed in a metal, rubber, or plastic sheath is called a **cable** (photo, page opposite). The sheath protects the wires from damage. Metal conduit also protects wires, but it is not considered a cable.

Individual wires are covered with rubber or plastic vinyl insulation. An exception is a bare copper grounding wire, which does not need an insulating cover. The insulation is color coded (chart, right) to identify the wire as a hot wire, a neutral wire, or a grounding wire.

In most wiring systems installed after 1965, the wires and cables are insulated with plastic vinyl. This type of insulation is very durable, and can last as long as the house itself.

Before 1965, wires and cables were insulated with rubber. Rubber insulation has a life expectancy of about 25 years. Old insulation that is cracked or damaged can be reinforced temporarily by wrapping the wire with plastic electrical tape. However, old wiring with cracked or damaged insulation should be inspected by a qualified electrician to make sure it is safe.

Wires must be large enough for the amperage rating of the circuit (chart, right). A wire that is too small can become dangerously hot. Wire sizes are categorized according to the American Wire Gauge (AWG) system. To check the size of a wire, use the wire stripper openings of a combination tool (page 14) as a guide.

Everything You Need:

Tools: cable ripper, combination tool, screwdriver, needlenose pliers.
Materials: wire nuts, pigtail wires (if needed).

Wire Color Chart

Wire color	Function
White	Neutral wire carrying current at zero voltage.
Black	Hot wire carrying current at full voltage.
Red	Hot wire carrying current at full voltage.
White, black markings	Hot wire carrying current at full voltage.
Green	Serves as a grounding pathway.
Bare copper	Serves as a grounding pathway.

Individual wires are color coded to identify their function. In some circuit installations, the white wire serves as a hot wire that carries voltage. If so, this white wire may be labeled with black tape or paint to identify it as a hot wire.

Wire Size Chart

Wire gauge	Wire capacity & use
#6	60 amps, 240 volts; central air conditioner, electric furnace.
#8	40 amps, 240 volts; electric range, central air conditioner.
#10	30 amps, 240 volts; window air conditioner, clothes dryer.
#12	20 amps, 120 volts; light fixtures, receptacles, microwave oven.
#14	15 amps, 120 volts; light fixtures, receptacles.
#16	Light-duty extension cords.
#18 to 22	Thermostats, doorbells, security systems.

Wire sizes (shown actual size) are categorized by the American Wire Gauge system. The larger the wire size, the smaller the AWG number.

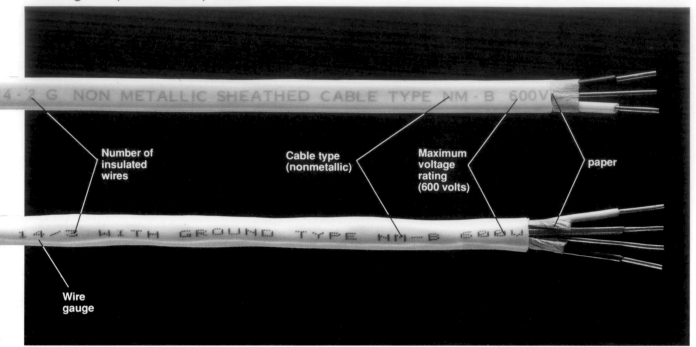

NM (nonmetallic) cable is labeled with the number of insulated wires it contains. The bare grounding wire is not counted. For example, a cable marked 14/2 G (or 14/2 WITH GROUND) contains two insulated 14-gauge wires, plus a bare copper grounding wire.

Cable marked 14/3 WITH GROUND has three 14-gauge wires plus a grounding wire. A strip of paper inside the cable protects the individual wires. NM cable also is stamped with a maximum voltage rating, as determined by Underwriters Laboratories (UL).

Aluminum Wire

Inexpensive aluminum wire was used in place of copper in many wiring systems installed during the late 1960s and early 1970s, when copper prices were high. Aluminum wire is identified by its silver color, and by the AL stamp on the cable sheathing. A variation, copper-clad aluminum wire, has a thin coating of copper bonded to a solid aluminum core.

By the early 1970s, all-aluminum wire was found to pose a safety hazard if connected to a switch or receptacle with brass or copper screw terminals. Because aluminum expands and contracts at a different rate than copper or brass, the wire connections could become loose. In some instances, fires resulted.

Existing aluminum wiring in homes is considered safe if proper installation methods have been followed, and if the wires are connected to special switches and receptacles designed to be used with aluminum wire. If you have aluminum wire in your home, have a qualified electrical inspector review the system. Copper-coated aluminum wire is not a hazard.

For a short while, switches and receptacles with an Underwriters Laboratories (UL) wire compatibility rating of AL-CU were used with both aluminum and copper wiring. However, these devices proved to be hazardous when connected to aluminum wire. AL-CU devices should not be used with aluminum wiring.

In 1971, switches and receptacles designed for use with aluminum wiring were introduced. They are marked CO/ALR. This mark is now the only approved rating for aluminum wires. If your home has aluminum wires connected to a switch or receptacle without a CO/ALR rating stamp, replace the device with a switch or receptacle rated CO/ALR.

A switch or receptacle that has no wire compatibility rating printed on the mounting strap or casing should not be used with aluminum wires. These devices are designed for use with copper wires only.

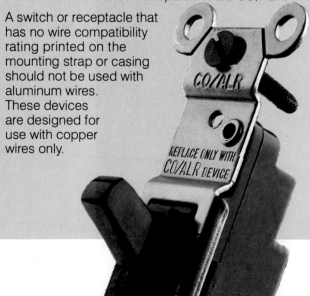

How to Strip NM (Nonmetallic) Cable & Wires

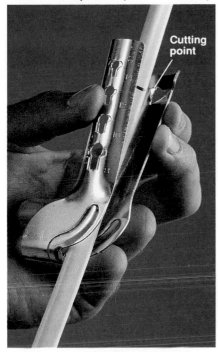

Cutting point

1 Measure and mark the cable 8" to 10" from end. Slide the cable ripper onto the cable, and squeeze tool firmly to force cutting point through plastic sheathing.

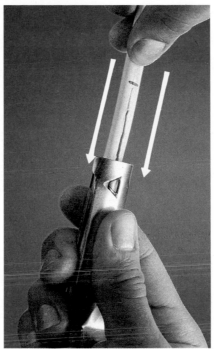

2 Grip the cable tightly with one hand, and pull the cable ripper toward the end of the cable to cut open the plastic sheathing.

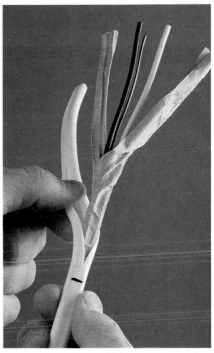

3 Peel back the plastic sheathing and the paper wrapping from the individual wires.

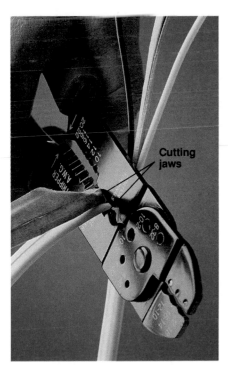

Cutting jaws

4 Cut away the excess plastic sheathing and paper wrapping, using the cutting jaws of a combination tool.

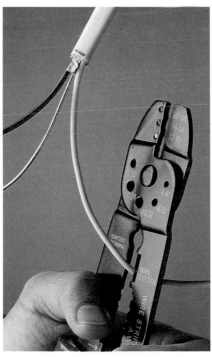

5 Cut the individual wires, if necessary, using the cutting jaws of the combination tool.

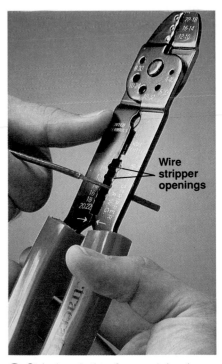

Wire stripper openings

6 Strip insulation from each wire, using the stripper openings. Choose the opening that matches the gauge of the wire, and take care not to nick or scratch the ends of the wires.

How to Connect Wires to Screw Terminals

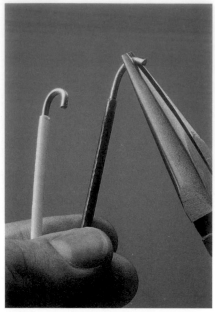

1 Strip about ¾" of insulation from each wire, using a combination tool. Choose the stripper opening that matches the gauge of the wire, then clamp wire in tool. Pull the wire firmly to remove plastic insulation.

2 Form a C-shaped loop in the end of each wire, using a needlenose pliers. The wire should have no scratches or nicks.

3 Hook each wire around the screw terminal so it forms a clockwise loop. Tighten screw firmly. Insulation should just touch head of screw. Never place the ends of two wires under a single screw terminal. Instead, use a pigtail wire (page opposite).

How to Connect Wires with Push-in Fittings

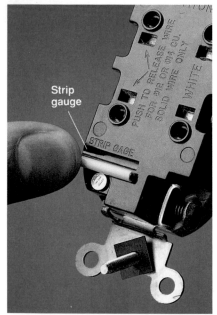

1 Mark the amount of insulation to be stripped from each wire, using the strip gauge on the back of the switch or receptacle. Strip the wires using a combination tool (step 1, above). Never use push-in fittings with aluminum wiring.

2 Insert the bare copper wires firmly into the push-in fittings on the back of the switch or receptacle. When inserted, wires should have no bare copper exposed.

Remove a wire from a push-in fitting by inserting a small nail or screwdriver in the release opening next to the wire. Wire will pull out easily.

How to Connect Two or More Wires with a Wire Nut

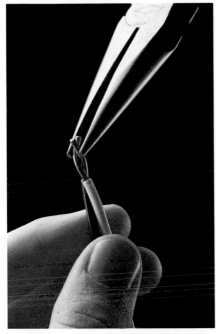

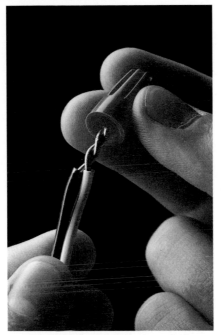

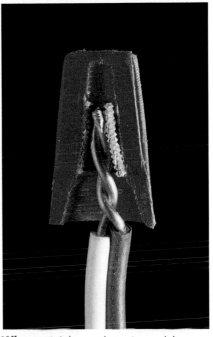

1 Strip about ½" of insulation from each wire. Hold the wires parallel, and twist them together in a clockwise direction, using a needle-nose pliers or combination tool.

2 Screw the wire nut onto the twisted wires. Tug gently on each wire to make sure it is secure. In a proper connection, no bare wire should be exposed past the bottom of the wire nut.

Wire nut (shown in cutaway) has metal threads that grip the bare ends of the wires. When connected, the wire nut should completely cover the bare wires.

How to Pigtail Two or More Wires

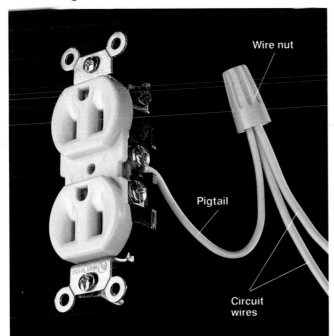

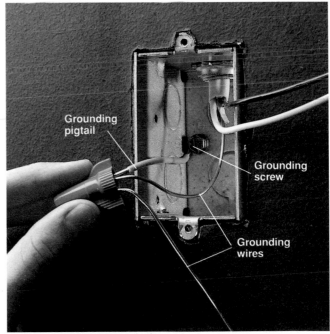

Connect two or more wires to a single screw termi-nal with a pigtail. A pigtail is a short piece of wire. One end of the pigtail connects to a screw terminal, and the other end connects to circuit wires, using a wire nut. A pigtail can also be used to lengthen cir-cuit wires that are too short.

Grounding pigtail has green insulation, and is avail-able with a preattached grounding screw. This ground-ing screw connects to the grounded metal electrical box. The end of the pigtail wire connects to the bare copper grounding wires with a wire nut.

Service Panels

Every home has a main service panel that distributes electrical current to the individual circuits. The main service panel usually is found in the basement, garage, or utility area, and can be identified by its metal casing. Before making any repair to your electrical system, you must shut off power to the correct circuit at the main service panel. The service panel should be indexed so circuits can be identified easily.

Service panels vary in appearance, depending on the age of the system. Very old wiring may operate on 30-amp service that has only two circuits. New homes can have 200-amp service with 30 or more circuits. Find the size of the service by reading the amperage rating printed on the main fuse block or main circuit breaker.

Regardless of age, all service panels have **fuses** or **circuit breakers** (pages 24 to 25) that control each circuit and protect them from overloads. In general, older service panels use fuses, while newer service panels use circuit breakers.

In addition to the main service panel, your electrical system may have a subpanel that controls some of the circuits in the home. A subpanel has its own circuit breakers or fuses, and is installed to control circuits that have been added to an existing wiring system.

The subpanel resembles the main service panel, but is usually smaller. It may be located near the main panel, or it may be found near the areas served by the new circuits. Garages and attics that have been updated often have their own subpanels. If your home has a subpanel, make sure that its circuits are indexed correctly.

When handling fuses or circuit breakers, make sure the area around the service panel is dry. Never remove the protective cover on the service panel. After turning off a circuit to make electrical repairs, remember to always test the circuit for power before touching any wires.

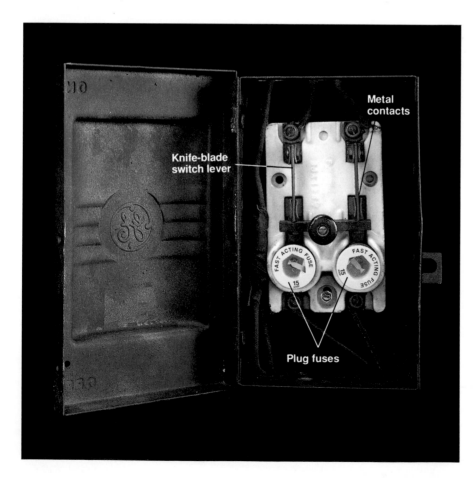

Knife-blade switch lever

Metal contacts

Plug fuses

A 30-amp service panel, common in systems installed before 1950, is identified by a ceramic fuse holder containing two plug fuses and a "knife-blade" switch lever. The fuse holder sometimes is contained in a black metal box mounted in an entryway or basement. This 30-amp service panel provides only 120 volts of power, and now is considered inadequate. For example, most home loan programs, like the FHA (Federal Housing Administration), require that 30-amp service be updated to 100 amps or more before a home can qualify for mortgage financing.

To shut off power to individual circuits in a 30-amp panel, carefully unscrew the plug fuses, touching only the insulated rim of the fuse. To shut off power to the entire house, open the knife-blade switch. Be careful not to touch the metal contacts on the switch.

A 60-amp fuse panel often is found in wiring systems installed between 1950 and 1965. It usually is housed in a gray metal cabinet that contains four individual plug fuses, plus one or two pull-out fuse blocks that hold cartridge fuses. This type of panel is regarded as adequate for a small, 1100-square-foot house that has no more than one 240-volt appliance. Many homeowners update 60-amp service to 100 amps or more so that additional lighting and appliance circuits can be added to the system. Home loan programs also may require that 60-amp service be updated before a home can qualify for financing.

To shut off power to a circuit, carefully unscrew the plug fuse, touching only its insulated rim. To shut off power to the entire house, hold the handle of the main fuse block and pull sharply to remove it. Major appliance circuits are controlled with another cartridge fuse block. Shut off the appliance circuit by pulling out this fuse block.

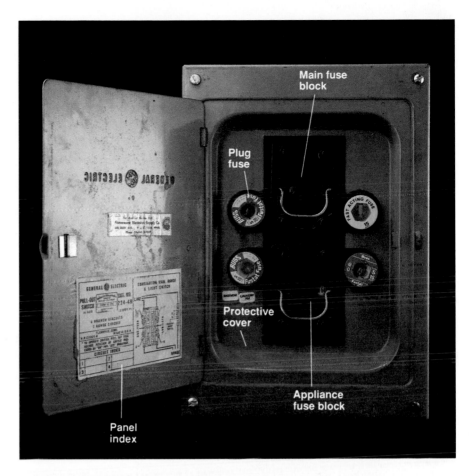

A circuit breaker panel providing 100 amps or more of power is common in wiring systems installed during the 1960s and later. A circuit breaker panel is housed in a gray metal cabinet that contains two rows of individual circuit breakers. The size of the service can be identified by reading the amperage rating of the main circuit breaker, which is located at the top of the main service panel.

A 100-amp service panel is now the minimum standard for most new housing. It is considered adequate for a medium-sized house with no more than three major electric appliances. However, larger houses with more electrical appliances require a service panel that provides 150 amps or more.

To shut off power to individual circuits in a circuit breaker panel, flip the lever on the appropriate circuit breaker to the OFF position. To shut off the power to the entire house, turn the main circuit breaker to the OFF position.

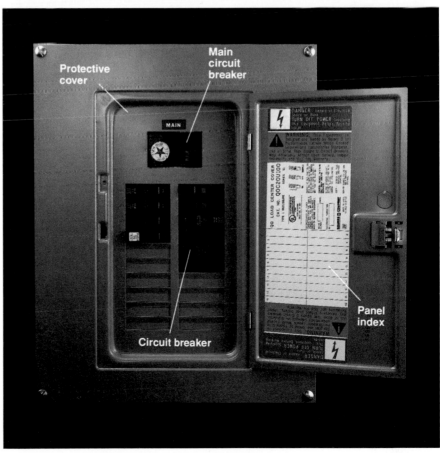

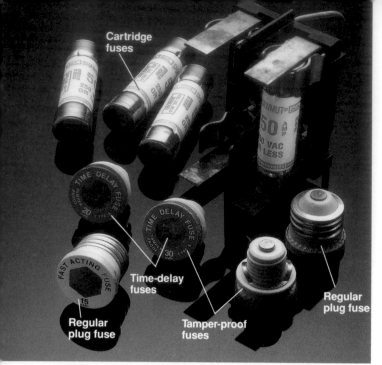

Cartridge fuses

Time-delay fuses

Regular plug fuse

Fast Acting Fuse

Regular plug fuse

Tamper-proof fuses

Regular plug fuse

Fuses are used in older service panels. Plug fuses usually control 120-volt circuits rated for 15, 20 or 30 amps. Tamper-proof plug fuses have threads that fit only matching sockets, making it impossible to install a wrong-sized fuse. Time-delay fuses absorb temporary heavy power load without blowing. Cartridge fuses control 240-volt circuits and range from 30 to 100 amps.

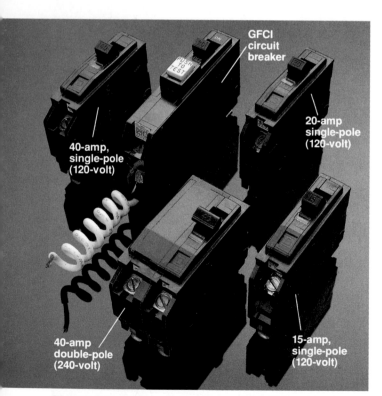

GFCI circuit breaker

20-amp single-pole (120-volt)

40-amp, single-pole (120-volt)

40-amp double-pole (240-volt)

15-amp, single-pole (120-volt)

Circuit breakers are used in newer service panels. Single-pole breakers rated for 15 or 20 amps control 120-volt circuits. Double-pole breakers rated for 20 to 50 amps control 240-volt circuits. GFCI (ground-fault circuit-interrupter) breakers provide shock protection for the entire circuit.

Fuses & Circuit Breakers

Fuses and circuit breakers are safety devices designed to protect the electrical system from short circuits and overloads. Fuses and circuit breakers are located in the main service panel.

Most service panels installed before 1965 rely on fuses to control and protect individual circuits. Screw-in plug fuses protect 120-volt circuits that power lights and receptacles. Cartridge fuses protect 240-volt appliance circuits and the main shutoff of the service panel.

Inside each fuse is a current-carrying metal alloy ribbon. If a circuit is overloaded (pages 26 to 27), the metal ribbon melts and stops the flow of power. A fuse must match the amperage rating of the circuit. Never replace a fuse with one that has a larger amperage rating.

In most service panels installed after 1965, circuit breakers protect and control individual circuits. Single-pole circuit breakers protect 120-volt circuits, and double-pole circuit breakers protect 240-volt circuits. Amperage ratings for circuit breakers range from 15 to 100 amps.

Each circuit breaker has a permanent metal strip that heats up and bends when voltage passes through it. If a circuit is overloaded, the metal strip inside the breaker bends enough to "trip" the switch and stop the flow of power. If a circuit breaker trips frequently even though the power demand is small, the mechanism inside the breaker may be worn out. Worn circuit breakers should be replaced by an electrician.

When a fuse blows or a circuit breaker trips, it is usually because there are too many light fixtures and plug-in appliances drawing power through the circuit. Move some of the plug-in appliances to another circuit, then replace the fuse or reset the breaker. If the fuse blows or breaker trips again immediately, there may be a short circuit in the system. Call a licensed electrician if you suspect a short circuit.

Everything You Need:

Tools: fuse puller (for cartridge fuses only).

Materials: replacement fuse.

How to Identify & Replace a Blown Plug Fuse

1 Go to the main service panel and locate the blown fuse. If the metal ribbon inside fuse is cleanly melted (right), the circuit was overloaded. If window in fuse is discolored (left), there was a short circuit in the system.

2 Unscrew the fuse, being careful to touch only the insulated rim of the fuse. Replace it with a fuse that has the same amperage rating.

How to Remove, Test & Replace a Cartridge Fuse

1 Remove cartridge fuses by gripping the handle of the fuse block and pulling sharply.

2 Remove the individual cartridge fuses from the block, using a fuse puller.

3 Test each fuse, using a continuity tester. If the tester glows, the fuse is good. If the tester does not glow, replace the fuse with one that has the same amperage rating.

How to Reset a Circuit Breaker

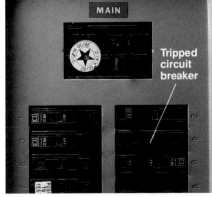

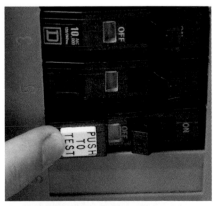

1 Open the service panel and locate the tripped breaker. The lever on the tripped breaker will be either in the OFF position, or in a position between ON and OFF.

2 Reset the tripped circuit breaker by pressing the circuit breaker lever all the way to the OFF position, then pressing it to the ON position.

Test GFCI circuit breakers by pushing TEST button. Breaker should trip to the OFF position. If not, the breaker is faulty and must be replaced by an electrician.

Evaluating Circuits for Safe Capacity

Every electrical circuit in a home has a "safe capacity." Safe capacity is the total amount of power the circuit wires can carry without tripping circuit breakers or blowing fuses. According to the National Electrical Code, the power used by light fixtures, lamps, tools, and appliances, called the "demand," must not exceed the safe capacity of the circuit.

Finding the safe capacity and the demand of a circuit is easy. Make these simple calculations to reduce the chances of tripped circuit breakers or blown fuses, or to help plan the location of new appliances or plug-in lamps.

First, determine the amperage and voltage rating of the circuit. If you have an up-to-date circuit map, these ratings should be indicated on the map. If not, open the service panel door and read the amperage rating printed on the circuit breaker, or on the rim of the fuse. The type of circuit breaker or fuse (page 24) indicates the voltage of the circuit.

Use the amperage and voltage ratings to find the safe capacity of the circuit. Safe capacities of the most common household circuits are given in the table at right.

Safe capacities can be calculated by multiplying the amperage rating by voltage. The answer is the total capacity, expressed in **watts**, a unit of electrical measurement. To find the safe capacity, reduce the total capacity by 20%.

Next, compare the safe capacity of the circuit to the total power demand. To find the demand, add the wattage ratings for all light fixtures, lamps, and appliances on the circuit. For lights, use the wattage rating printed on the light bulbs. Wattage ratings for appliances often are printed on the manufacturer's label. Approximate wattage ratings for many common household items are given in the table on the opposite page. If you are unsure about the wattage rating of a tool or appliance, use the highest number shown in the table to make calculations.

Compare the power demand to the safe capacity. The power demand should not exceed the safe capacity of the circuit. If it does, you must move lamps or appliances to another circuit. Or, make sure that the power demand of the lamps and appliances turned on **at the same time** does not exceed the safe capacity of the circuit.

Amps x Volts	Total Capacity	Safe Capacity
15 A x 120 V =	1800 watts	1440 watts
20 A x 120 V =	2400 watts	1920 watts
25 A x 120 V =	3000 watts	2400 watts
30 A x 120 V =	3600 watts	2880 watts
20 A x 240 V =	4800 watts	3840 watts
30 A x 240 V =	7200 watts	5760 watts

How to Find Wattage & Amperage Ratings

Light bulb wattage ratings are printed on the top of the bulb. If a light fixture has more than one bulb, remember to add the wattages of all the bulbs to find the total wattage of the fixture.

Appliance wattage ratings are often listed on the manufacturer's label. Or, use table of typical wattage ratings on the opposite page.

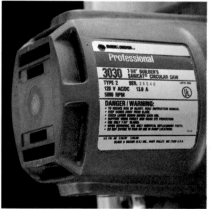

Amperage rating can be used to find the wattage of an appliance. Multiply the amperage by the voltage of the circuit. For example, a 13-amp, 120-volt circular saw is rated for 1560 watts.

Sample Circuit Evaluation

Circuit # _6_ **Amps** _20_ **Volts** _120_ **Total capacity** _2400_ (watts) **Safe capacity** _1920_ (watts)

Appliance or fixture	Notes	Wattage rating
REFRIGERATOR	CONSTANT USE	480
CEILING LIGHT	3 - 60 WATT BULBS	180
MICROWAVE OVEN		625
ELECTRIC CAN OPENER	OCCASIONAL USE	144
STEREO	PORTABLE BOOM BOX	300
CEILING LIGHT (HALLWAY)	2 60 WATT BULBS	120
	Total demand:	1849 (watts)

Photocopy this sample circuit evaluation to keep a record of the power demand of each circuit. The words and numbers printed in blue will not reproduce on photocopies. In this sample kitchen circuit, the demand on the circuit is very close to the safe capacity. Adding another appliance, such as an electric frying pan, could overload the circuit and cause a fuse to blow or a circuit breaker to trip.

Typical Wattage Ratings (120-volt Circuit Except Where Noted)

Appliance	Amps	Watts	Appliance	Amps	Watts
Air conditioner (central)	21 (240-v)	5040	Hair dryer	5 to 10	600 to 1200
Air conditioner (window)	6 to 13	720 to 1560	Heater (portable)	7 to 12	840 to 1440
Blender	2 to 4	240 to 480	Microwave oven	4 to 7	480 to 840
Broiler	12.5	1500	Range (oven/stove)	16 to 32 (240-v)	3840 to 7680
Can opener	1.2	144	Refrigerator	2 to 4	240 to 480
Circular saw	10 to 12	1200 to 1440	Router	8	960
Coffee maker	4 to 8	480 to 960	Sander (portable)	2 to 5	240 to 600
Clothes dryer	16.5 to 34 (240-v)	3960 to 8160	Saw (table)	7 to 10	840 to 1200
Clothes iron	9	1080	Sewing machine	1	120
Computer	4 to 7	480 to 840	Stereo	2.5 to 4	300 to 480
Dishwasher	8.5 to 12.5	1020 to 1500	Television (b & w)	2	240
Drill (portable)	2 to 4	240 to 480	Television (color)	2.5	300
Fan (ceiling)	3.5	420	Toaster	9	1080
Fan (portable)	2	240	Trash compactor	4 to 8	480 to 960
Freezer	2 to 4	240 to 480	Vacuum cleaner	6 to 11	720 to 1320
Frying pan	9	1080	Waffle iron	7.5	900
Furnace, forced-air gas	6.5 to 13	780 to 1560	Washing machine	12.5	1500
Garbage disposer	3.5 to 7.5	420 to 900	Water heater	10.5 to 21 (240-v)	2520 to 5040

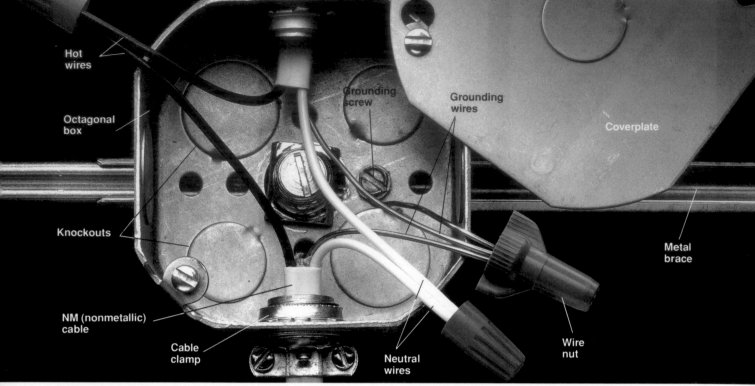

Octagonal boxes usually contain wire connections for ceiling fixtures. Cables are inserted into the box through knockout openings, and are held with cable clamps. Because the ceiling fixture attaches directly to the box, the box should be anchored firmly to a framing member. Often, it is nailed directly to a ceiling joist. However, metal braces are available that allow a box to be mounted between joists or studs. A properly installed octagonal box can support a ceiling fixture weighing up to 35 pounds. Any box must be covered with a tightly fitting coverplate, and the box must not have open knockouts.

Electrical Boxes

The National Electrical Code requires that wire connections or cable splices be contained inside an approved metal or plastic box. This shields framing members and other flammable materials from electrical sparks. If you have exposed wire connections or cable splices, protect your home by installing electrical boxes.

Electrical boxes come in several shapes. Rectangular and square boxes are used for switches and receptacles. Rectangular (2" × 3") boxes are used for single switches or duplex receptacles. Square (4" × 4") boxes are used anytime it is convenient for two switches or receptacles to be wired or "ganged" in one box, an arrangement common in kitchens or entry hallways. Octagonal electrical boxes contain wire connections for ceiling fixtures.

All electrical boxes are available in different depths. A box must be deep enough so a switch or receptacle can be removed or installed easily without crimping and damaging the circuit wires. Replace an undersized box with a larger box, using the Electrical Box Chart (below) as a guide. The NEC also says that all electrical boxes must remain accessible. Never cover an electrical box with drywall, paneling, or wallcoverings.

Electrical Box Chart

Box Shape		Maximum number of individual wires in box*	
		14-gauge	12-gauge
2"×3" rectangular	2½" deep	3	3
	3½" deep	5	4
4"×4" square	1½" deep	6	5
	2⅛" deep	9	7
Octagonal	1½" deep	4	3
	2⅛" deep	7	6

* Do not count pigtail wires or grounding wires.

Common Electrical Boxes

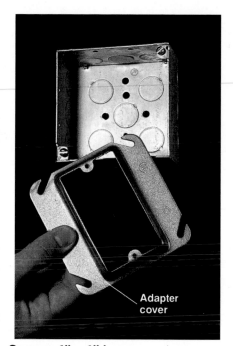

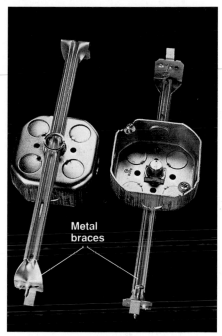

Rectangular boxes are used with wall switches and duplex receptacles. Single-size rectangular boxes (shown above) may have detachable sides that allow them to be ganged together to form double-size boxes.

Square 4" × 4" boxes are large enough for most wiring applications. They are used for cable splices and ganged receptacles or switches. To install one switch or receptacle in a square box, use an adapter cover.

Braced octagonal boxes fit between ceiling joists. The metal braces extend to fit any joist spacing, and are nailed or screwed to framing members.

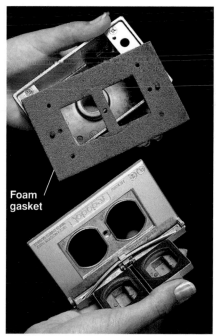

Outdoor boxes have sealed seams and foam gaskets to guard a switch or receptacle against moisture. Corrosion-resistant coatings protect all metal parts.

Retrofit boxes upgrade older boxes to larger sizes. One type (above) has built-in clamps that tighten against the inside of a wall and hold the box in place. A retrofit box with flexible brackets is shown on pages 32 to 33.

Plastic boxes are common in new construction. They can be used only with NM (nonmetallic) cable. Box may include preattached nails for anchoring the box to framing members.

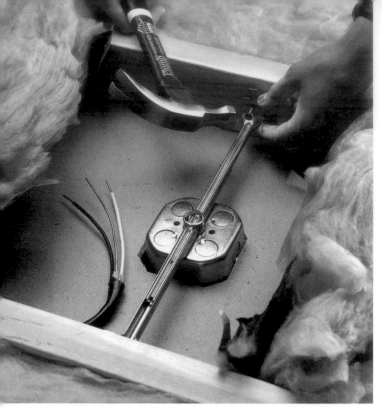

Electrical boxes are required for all wire connections. The box protects wood and other flammable materials from electrical sparks (arcing). Electrical boxes should always be anchored to joists or studs.

Installing an Electrical Box

Install an electrical box any time you find exposed wire connections or cable splices. Exposed connections sometimes can be found in older homes, where wires attach to light fixtures. Exposed splices can be found in areas where NM (nonmetallic) cable runs through uncovered joists or wall studs, such as in an unfinished basement or utility room.

When installing an electrical box, make sure there is enough cable to provide about 8" of wire inside the box. If the wires are too short, you can add pigtails to lengthen them. If the electrical box is metal, make sure the circuit grounding wires are pigtailed to the box.

Everything You Need:

Tools: neon circuit tester, screwdriver, hammer, combination tool.

Materials: screws or nails, electrical box, cable connectors, pigtail wire, wire nuts.

How to Install an Electrical Box for Cable Splices

1 Turn off power to circuit wires at the main service panel. Carefully remove any tape or wire nuts from the exposed splice. Avoid contact with the bare wire ends until the wires have been tested for power.

2 Test for power. Touch one probe of a circuit tester to the black hot wires, and touch other probe to the white neutral wires. The tester should not glow. If it does, the wires are still hot. Shut off power to correct circuit at the main service panel. Disconnect the splice wires.

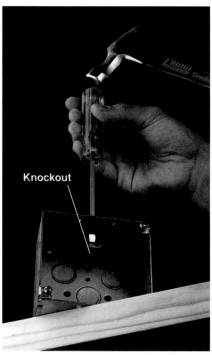

Knockout

3 Open one knockout for each cable that will enter the box, using a hammer and screwdriver. Any unopened knockouts should remain sealed.

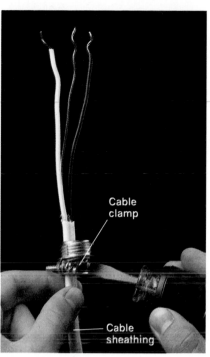

4 Anchor the electrical box to a wooden framing member, using screws or nails.

5 Thread each cable through a cable clamp. Tighten the clamp with a screwdriver. Do not overtighten. Overtightening can damage cable sheathing.

6 Insert the cables into the electrical box, and screw a locknut onto each cable clamp.

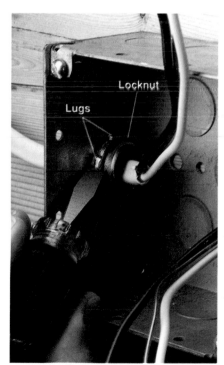

7 Tighten the locknuts by pushing against the lugs with the blade of a screwdriver.

8 Use wire nuts to reconnect the wires. Pigtail the copper grounding wires to the green grounding screw in the back of the box.

9 Carefully tuck the wires into the box, and attach the coverplate. Turn on the power to the circuit at the main service panel. Make sure the box remains accessible, and is not covered with finished walls or ceilings.

Replacing an Electrical Box

Replace any electrical box that is too small for the number of wires it contains. Forcing wires into an undersized box can damage wires, disturb wire connections, and create a potential fire hazard.

Boxes that are too small often are found when repairing or replacing switches, receptacles, or light fixtures. If you find a box so small that you have difficulty fitting the wires inside, replace it with a larger box. Use the chart on page 28 as a guide when choosing a replacement box.

Metal and plastic retrofit electrical boxes are available in a variety of styles, and can be purchased at any hardware store or home center. Most can be installed without damaging the wall surfaces.

Everything You Need:

Tools: screwdriver, neon circuit tester, reciprocating saw, hammer, needlenose pliers.

Materials: electrical tape, retrofit electrical box with flexible brackets, grounding screw.

How to Replace an Electrical Box

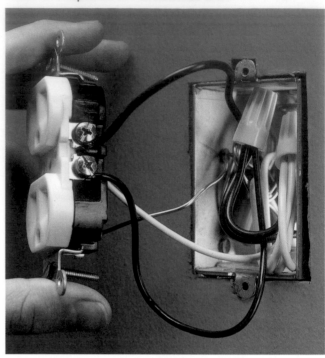

1 Shut off the power to the circuit at the main service panel. Test for power with a neon circuit tester (switches, page 47; receptacles, page 58; light fixtures, page 66). Disconnect and remove receptacle, switch, or fixture from the existing box.

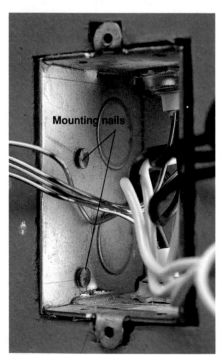

2 Examine the box to determine how it was installed. Most older metal boxes are attached to framing members with nails, and the nail heads will be visible inside the box.

3 Cut through the mounting nails by slipping the metal-cutting blade of a reciprocating saw between the box and framing member. Take care not to damage the circuit wires. Disconnect wires.

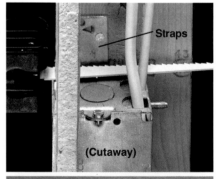

If box is mounted with straps (shown with wall cut away) remove the box by cutting through the straps, using a reciprocating saw and metal-cutting blade. Take care not to damage the wires.

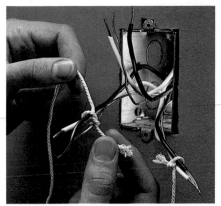

4 To prevent wires from falling into wall cavity, gather wires from each cable and tie them together, using pieces of string.

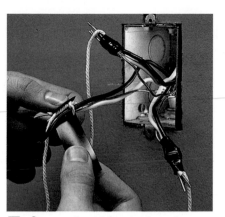

5 Secure the string to the wires, using a piece of plastic electrical tape.

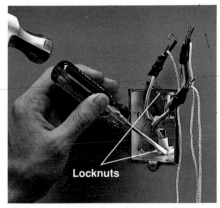

Locknuts

6 Disconnect the internal clamps or locknuts that hold the circuit cables to the box.

7 Pull old electrical box from wall. Take care not to damage insulation on circuit wires, and hold on to string to make sure wires do not fall inside wall cavity.

8 Tape the wires to the edge of the wall cutout.

9 Punch out one knockout for each cable that will enter the new electrical box, using a screwdriver and hammer.

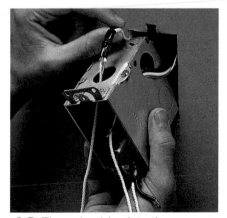

10 Thread cables into the new electrical box, and slide the box into the wall opening. Tighten the internal clamps or locknuts holding the circuit cables to the electrical box. Remove the strings.

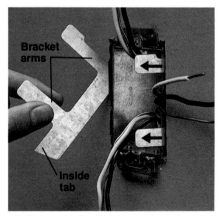

Bracket arms

Inside tab

11 Insert flexible brackets into the wall on each side of the electrical box. Pull out bracket arms until inside tab is tight against inside of the wall.

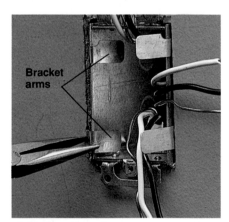

Bracket arms

12 Bend bracket arms around walls of box, using needlenose pliers. Reinstall the fixture, and turn on the power to the circuit at the main service panel.

Rotary snap switches are found in many installations completed between 1900 and 1920. Handle is twisted clockwise to turn light on and off. The switch is enclosed in a ceramic housing.

Push-button switches were widely used from 1920 until about 1940. Many switches of this type are still in operation. Reproductions of this switch type are available for restoration projects.

Toggle switches were introduced in the 1930s. This early design has a switch mechanism that is mounted in a ceramic housing sealed with a layer of insulating paper.

Common Wall-switch Problems

An average wall switch is turned on and off more than 1,000 times each year. Because switches receive constant use, wire connections can loosen and switch parts gradually wear out. If a switch no longer operates smoothly, it must be repaired or replaced.

The methods for repairing or replacing a switch vary slightly, depending on the switch type and its location along an electrical circuit. When working on a switch, use the photographs on pages 36 to 41 to identify your switch type and its wiring configuration. Individual switch styles may vary from manufacturer to manufacturer, but the basic switch types are universal.

It is possible to replace most ordinary wall switches with a specialty switch, like a timer switch or an electronic switch. When installing a specialty switch (page 41), make sure it is compatible with the wiring configuration of the switch box.

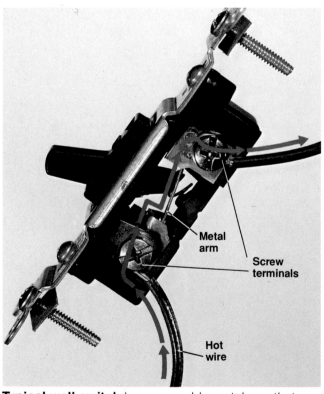

Metal arm

Screw terminals

Hot wire

Typical wall switch has a movable metal arm that opens and closes the electrical circuit. When the switch is ON, the arm completes the circuit and power flows between the screw terminals and through the black hot wire to the light fixture. When the switch is OFF, the arm lifts away to interrupt the circuit, and no power flows. Switch problems can occur if the screw terminals are not tight, or if the metal arm inside the switch wears out.

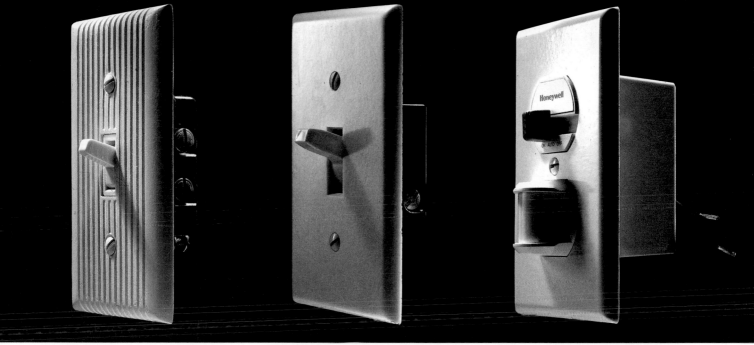

Toggle switches were improved during the 1950s, and are now the most commonly used type. This switch type was the first to use a sealed plastic housing that protects the inner switch mechanism from dust and moisture.

Mercury switches became common in the early 1960s. They conduct electrical current by means of a sealed vial of mercury. Although more expensive than other types, mercury switches are durable: some are guaranteed for 50 years.

Electronic motion-sensor switch has an infrared eye that senses movement and automatically turns on lights when a person enters a room. Motion-sensor switches can provide added security against intruders.

Problem	Repair
Fuse burns out or circuit breaker trips when the switch is turned on.	1. Tighten any loose wire connections on switch (pages 46 to 47). 2. Move lamps or plug-in appliances to other circuits to prevent overloads (page 26). 3. Test switch (pages 42 to 45), and replace, if needed (pages 46 to 49). 4. Repair or replace faulty fixture (pages 64 to 76) or faulty appliance.
Light fixture or permanently installed appliance does not work.	1. Replace burned-out light bulb. 2. Check for blown fuse or tripped circuit breaker to make sure circuit is operating (pages 24 to 25). 3. Check for loose wire connections on switch (pages 46 to 47). 4. Test switch (pages 42 to 45), and replace, if needed (pages 46 to 49). 5. Repair or replace light fixture (pages 64 to 76) or appliance.
Light fixture flickers.	1. Tighten light bulb in the socket. 2. Check for loose wire connections on switch (pages 46 to 47). 3. Repair or replace light bulb fixture (pages 64 to 76), or switch (pages 46 to 49).
Switch buzzes or is warm to the touch.	1. Check for loose wire connections on switch (pages 46 to 47). 2. Test switch (pages 42 to 45), and replace, if needed (pages 46 to 49). 3. Move lamps or appliances to other circuits to reduce demand (page 26).
Switch lever does not stay in position.	Replace worn-out switch (pages 46 to 49).

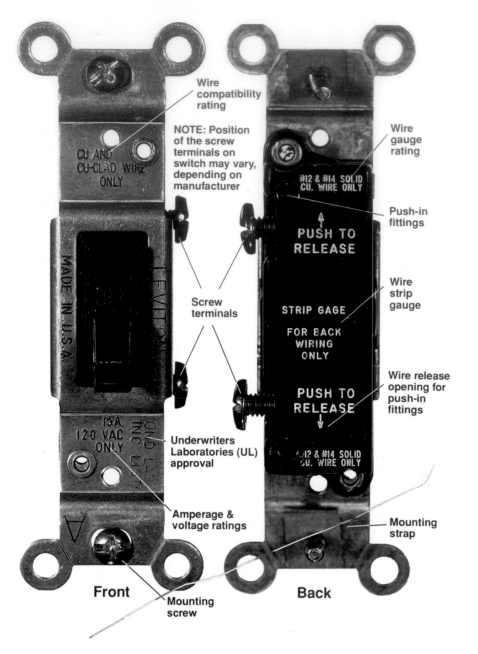

Wire compatibility rating

NOTE: Position of the screw terminals on switch may vary, depending on manufacturer

CU AND CU-CLAD WIRE ONLY

MADE IN U.S.A.

LEVITON

Screw terminals

15A. 120 VAC ONLY
UND. LAB. INC.

Underwriters Laboratories (UL) approval

Amperage & voltage ratings

Front

Mounting screw

Wire gauge rating

#12 & #14 SOLID CU. WIRE ONLY

PUSH TO RELEASE

STRIP GAGE FOR BACK WIRING ONLY

PUSH TO RELEASE

#12 & #14 SOLID CU. WIRE ONLY

Push-in fittings

Wire strip gauge

Wire release opening for push-in fittings

Mounting strap

Back

Wall switches are available in three general types. To repair or replace a switch, it is important to identify its type.

Single-pole switches are used to control a set of lights from one location. Three-way switches are used to control a set of lights from two different locations, and are always installed in pairs. Four-way switches are used in combination with a pair of three-way switches to control a set of lights from three or more locations.

Identify switch types by counting the screw terminals. Single-pole switches have two screw terminals, three-way switches have three screw terminals, and four-way switches have four.

Some switches may include a grounding screw terminal, which is identified by its green color. Although grounded switches are not required by most electrical codes, some electricians recommend they be used in bathrooms, kitchens, and basements. When pigtailed to the grounding wires, the grounding screw provides added protection against shock.

When replacing a switch, choose a new switch that has the same number of screw terminals as the old one. The location of the screws on the switch body varies, depending on the manufacturer, but these differences will not affect the switch operation.

Newer switches may have push-in fittings in addition to the screw terminals. Some specialty switches have wire leads instead of screw terminals. They are connected to circuit wires with wire nuts.

A wall switch is connected to circuit wires with screw terminals or with push-in fittings on the back of the switch. A switch may have a stamped strip gauge that indicates how much insulation must be stripped from the circuit wires to make the connections.

The switch body is attached to a metal mounting strap that allows it to be mounted in an electrical box. Several rating stamps are found on the strap and on the back of the switch. The abbreviation UL or UND. LAB. INC. LIST means that the switch meets the safety standards of the Underwriters Laboratories. Switches also are stamped with maximum voltage and amperage ratings. Standard wall switches are rated 15A, 125V. Voltage ratings of 110, 120, and 125 are considered to be identical for purposes of identification.

For standard wall switch installations, choose a switch that has a wire gauge rating of #12 or #14. For wire systems with solid-core copper wiring, use only switches marked COPPER or CU. For aluminum wiring (page 18), use only switches marked CO/ALR. Switches marked AL/CL can no longer be used with aluminum wiring, according to the National Electrical Code.

Single-pole Wall Switches

A single-pole switch is the most common type of wall switch. It usually has ON-OFF markings on the switch lever, and is used to control a set of lights, an appliance, or a receptacle from a single location. A single-pole switch has two screw terminals. Some types also may have a grounding screw. When installing a single-pole switch, check to make sure the ON marking shows when the switch lever is in the up position.

In a correctly wired single-pole switch, a hot circuit wire is attached to each screw terminal. However, the color and number of wires inside the switch box will vary, depending on the location of the switch along the electrical circuit.

If two cables enter the box, then the switch lies in the middle of the circuit. In this installation, both of the hot wires attached to the switch are black.

If only one cable enters the box, then the switch lies at the end of the circuit. In this installation (sometimes called a switch loop), one of the hot wires is black, but the other hot wire usually is white. A white hot wire sometimes is coded with black tape or paint.

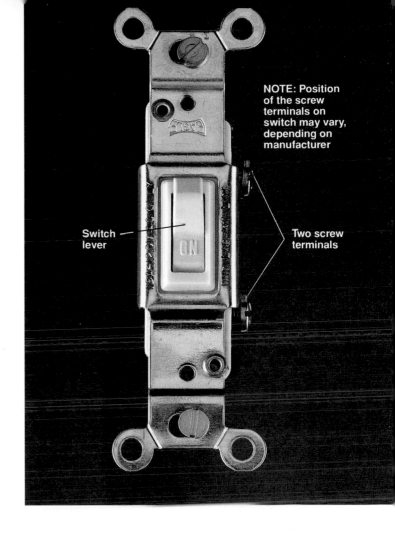

NOTE: Position of the screw terminals on switch may vary, depending on manufacturer

Switch lever

Two screw terminals

Typical Single-pole Switch Installations

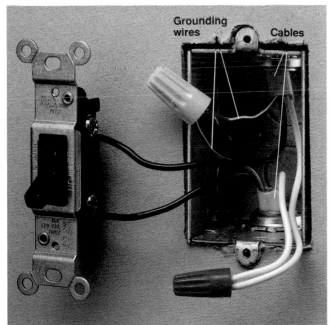

Grounding wires

Cables

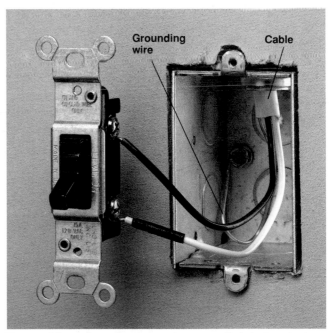

Grounding wire

Cable

Two cables enter the box when a switch is located in the middle of a circuit. Each cable has a white and a black insulated wire, plus a bare copper grounding wire. The black wires are hot, and are connected to the screw terminals on the switch. The white wires are neutral and are joined together with a wire nut. Grounding wires are pigtailed to the grounded box.

One cable enters the box when a switch is located at the end of a circuit. The cable has a white and a black insulated wire, plus a bare copper grounding wire. In this installation, both of the insulated wires are hot. The white wire may be labeled with black tape or paint to identify it as a hot wire. The grounding wire is connected to the grounded metal box.

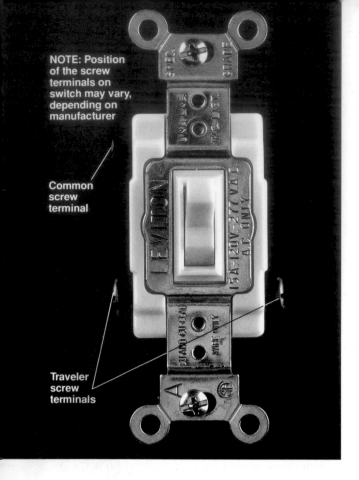

NOTE: Position of the screw terminals on switch may vary, depending on manufacturer

Common screw terminal

Traveler screw terminals

Three-way Wall Switches

Three-way switches have three screw terminals, and do not have ON-OFF markings. Three-way switches are always installed in pairs, and are used to control a set of lights from two locations.

One of the screw terminals on a three-way switch is darker than the others. This screw is the **common screw terminal**. The position of the common screw terminal on the switch body may vary, depending on the manufacturer. Before disconnecting a three-way switch, always label the wire that is connected to the common screw terminal. It must be reconnected to the common screw terminal on the new switch.

The two lighter-colored screw terminals on a three-way switch are called the **traveler screw terminals**. The traveler terminals are interchangeable, so there is no need to label the wires attached to them.

Because three-way switches are installed in pairs, it sometimes is difficult to determine which of the switches is causing a problem. The switch that receives greater use is more likely to fail, but you may need to inspect both switches to find the source of the problem.

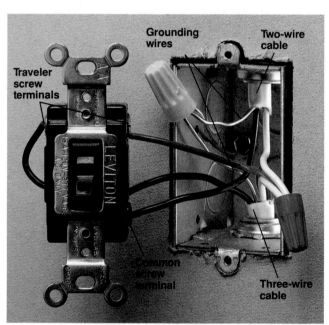

Traveler screw terminals

Grounding wires

Two-wire cable

Common screw terminal

Three-wire cable

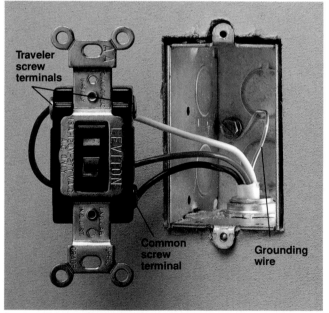

Traveler screw terminals

Common screw terminal

Grounding wire

Two cables enter box if the switch lies in the middle of a circuit. One cable has two wires, plus a bare copper grounding wire; the other cable has three wires, plus a ground. The black wire from the two-wire cable is connected to the dark, common screw terminal. The red and black wires from the three-wire cable are connected to the traveler screw terminals. The white neutral wires are joined together with a wire nut, and the grounding wires are pigtailed to the grounded metal box.

One cable enters the box if the switch lies at the end of the circuit. The cable has a black wire, red wire, and white wire, plus a bare copper grounding wire. The black wire must be connected to the common screw terminal, which is darker than the other two screw terminals. The white and red wires are connected to the two traveler screw terminals. The bare copper grounding wire is connected to the grounded metal box.

Four-way Wall Switches

Four-way switches have four screw terminals, and do not have ON-OFF markings. Four-way switches are always installed between a pair of three-way switches. This switch combination makes it possible to control a set of lights from three or more locations. Four-way switches are not common, but are sometimes found in homes that have very large rooms or long hallways. Switch problems in a four-way installation can be caused by loose connections or worn parts in a four-way switch or in one of the three-way switches (page opposite).

In a typical installation, two pairs of color-matched wires are connected to the four-way switch. To simplify installation, newer four-way switches have screw terminals that are paired by color. One pair of screws usually is copper, while the other pair is brass. When installing the switch, match the wires to the screw terminals by color. For example, if a red wire is connected to one of the brass screw terminals, make sure the other red wire is connected to the remaining brass screw terminal.

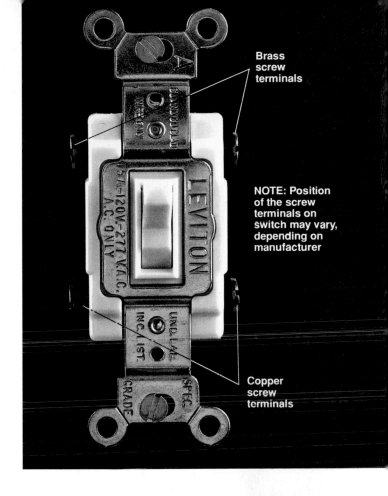

Brass screw terminals

NOTE: Position of the screw terminals on switch may vary, depending on manufacturer

Copper screw terminals

Typical Four-way Switch Installation

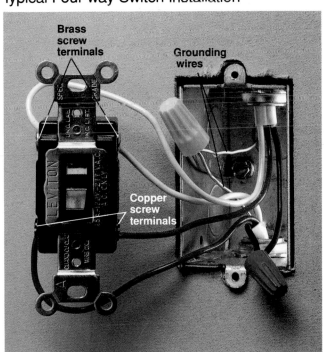

Brass screw terminals

Grounding wires

Copper screw terminals

Four wires are connected to a four-way switch. One pair of color-matched wires is connected to the copper screw terminals, while another pair is connected to the brass screw terminals. A third pair of wires are connected inside the box with a wire nut. The two bare copper grounding wires are pigtailed to the grounded metal box.

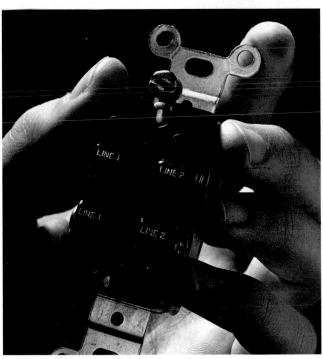

Switch variation: Some four-way switches have a wiring guide stamped on the back to help simplify installation. For the switch shown above, one pair of color-matched circuit wires will be connected to the screw terminals marked LINE 1, while the other pair of wires will be attached to the screw terminals marked LINE 2.

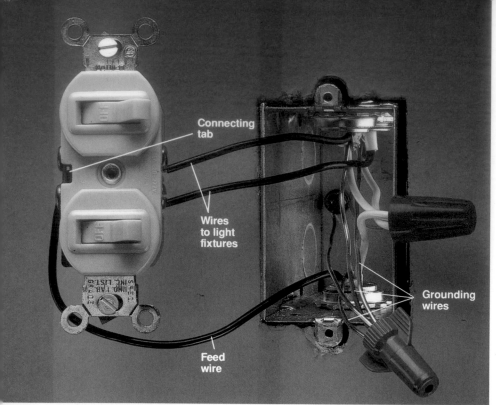

Single-circuit wiring: Three black wires are attached to the switch. The black feed wire bringing power into the box is connected to the side of the switch that has a connecting tab. The wires carrying power out to the light fixtures or appliances are connected to the side of the switch that **does not** have a connecting tab. The white neutral wires are connected together with a wire nut.

Double Switches

A double switch has two switch levers in a single housing. It is used to control two light fixtures or appliances from the same switch box.

In most installations, both halves of the switch are powered by the same circuit. In these **single-circuit** installations, three wires are connected to the double switch. One wire, called the "feed" wire, supplies power to both halves of the switch. The other wires carry power out to the individual light fixtures or appliances.

In rare installations, each half of the switch is powered by a separate circuit. In these **separate-circuit** installations, four wires are connected to the switch, and the metal connecting tab joining two of the screw terminals is removed (photo below).

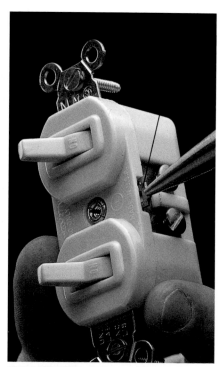

Separate-circuit wiring: Four black wires are attached to the switch. Feed wires from the power source are attached to the side of switch that has a connecting tab, and the connecting tab is removed (photo, right). Wires carrying power from the switch to light fixtures or appliances are connected to the side of the switch that **does not** have a connecting tab. White neutral wires are connected together with a wire nut.

Remove the connecting tab on a double switch when wired in a separate-circuit installation. The tab can be removed with needle-nose pliers or a screwdriver.

Pilot-light Switches

A pilot-light switch has a built-in bulb that glows when power flows through the switch to a light fixture or appliance. Pilot-light switches often are installed for convenience if a light fixture or appliance cannot be seen from the switch location. Basement lights, garage lights, and attic exhaust fans frequently are controlled by pilot-light switches.

A pilot-light switch requires a neutral wire connection. A switch box that contains a single two-wire cable has only hot wires, and cannot be fitted with a pilot-light switch.

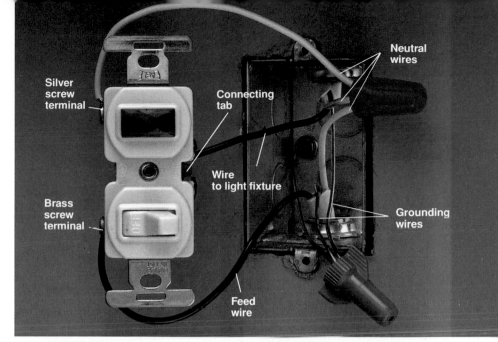

Pilot-light switch wiring: Three wires are connected to the switch. One black wire is the feed wire that brings power into the box. It is connected to the brass screw terminal on the side of the switch that **does not** have a connecting tab. The white neutral wires are pigtailed to the silver screw terminal. Black wire carrying power out to light fixture or appliance is connected to screw terminal on side of the switch that has a connecting tab.

Switch/receptacles

A switch/receptacle combines a grounded receptacle with a single-pole wall switch. In a room that does not have enough wall receptacles, electrical service can be improved by replacing a single-pole switch with a switch/receptacle.

A switch/receptacle requires a neutral wire connection. A switch box that contains a single two-wire cable has only hot wires, and cannot be fitted with a switch/receptacle.

A switch/receptacle can be installed in one of two ways. In the most common installations, the receptacle is hot even when the switch is off (photo, right).

In rare installations, a switch/-receptacle is wired so the receptacle is hot only when the switch is on. In this installation, the hot wires are reversed, so that the feed wire is attached to the brass screw terminal on the side of the switch that does not have a connecting tab.

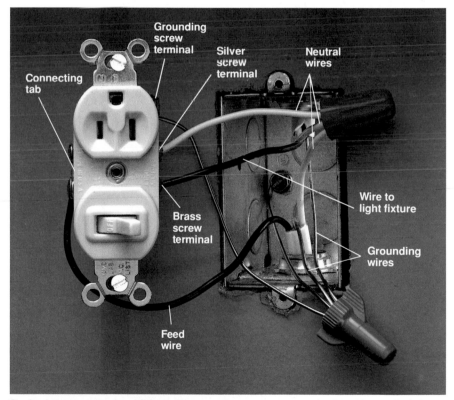

Switch/receptacle wiring: Three wires are connected to the switch/-receptacle. One of the hot wires is the feed wire that brings power into the box. It is connected to the side of the switch that has a connecting tab. The other hot wire carries power out to the light fixture or appliance. It is connected to the brass screw terminal on the side that **does not** have a connecting tab. The white neutral wire is pigtailed to the silver screw terminal. The grounding wires must be pigtailed to the green grounding screw on the switch/receptacle and to the grounded metal box.

Testing Switches for Continuity

A switch that does not work properly may have worn or broken internal parts. Test for internal wear with a battery-operated continuity tester. The continuity tester detects any break in the metal pathway inside the switch. Replace the switch if the continuity tester shows the switch to be faulty.

Never use a continuity tester on wires that might carry live current. Always shut off the power and disconnect the switch before testing for continuity.

Some specialty switches, like dimmers, cannot be tested for continuity. Electronic switches can be tested for manual operation using a continuity tester, but the automatic operation of these switches cannot be tested.

Everything You Need:

Tools: continuity tester.

How to Test a Single-pole Wall Switch

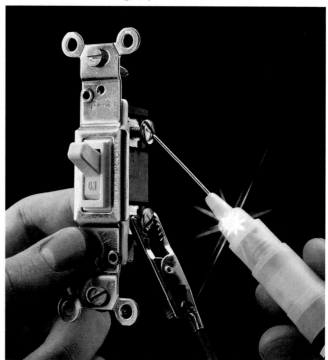

Attach clip of tester to one of the screw terminals. Touch the tester probe to the other screw terminal. Flip switch lever from ON to OFF. If switch is good, tester glows when lever is ON, but not when OFF.

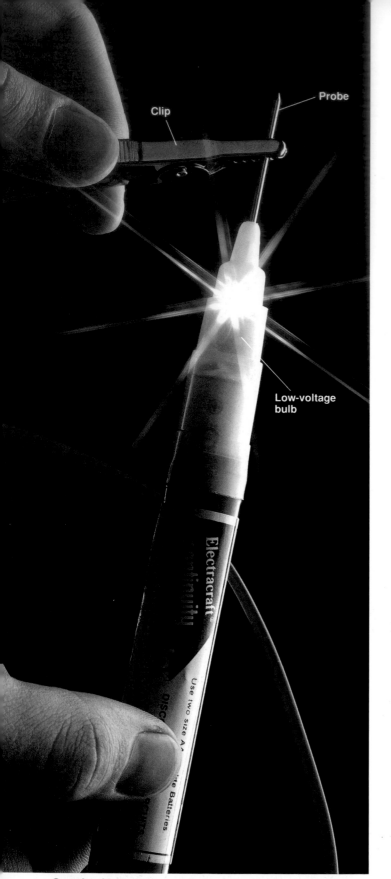

Clip

Probe

Low-voltage bulb

Continuity tester uses battery-generated current to test the metal pathways running through switches and other electrical fixtures. Always "test" the tester before use. Touch the tester clip to the metal probe. The tester should glow. If not, then the battery or light bulb is dead and must be replaced.

How to Test a Three-way Wall Switch

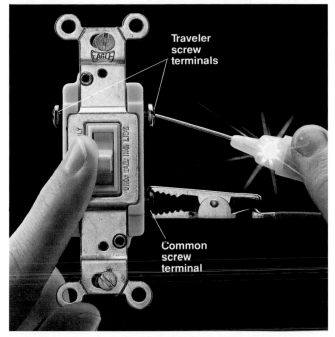

1 Attach tester clip to the dark common screw terminal. Touch the tester probe to one of the traveler screw terminals, and flip switch lever back and forth. If switch is good, the tester should glow when the lever is in one position, but not both.

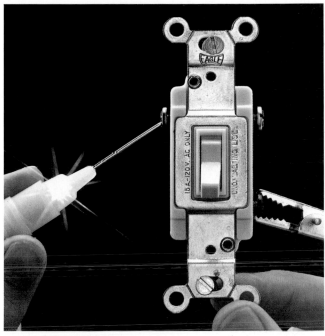

2 Touch probe to the other traveler screw terminal, and flip the switch lever back and forth. If switch is good, the tester will glow only when the switch lever is in the position opposite from the positive test in step 1.

How to Test a Four-way Wall Switch

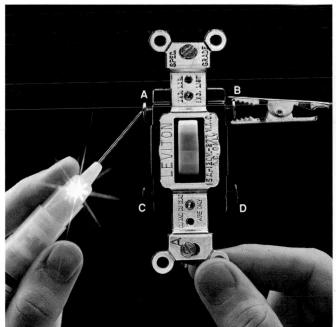

1 Test switch by touching probe and clip of continuity tester to each pair of screw terminals (A-B, C-D, A-D, B-C, A-C, B-D). The test should show continuous pathways between two different pairs of screw terminals. Flip lever to opposite position, and repeat test. Test should show continuous pathways between two different pairs of screw terminals.

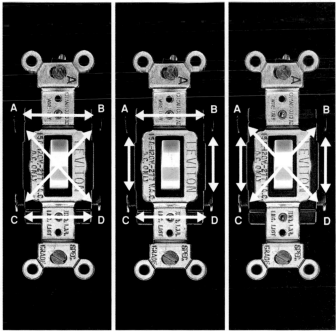

2 If switch is good, test will show a total of four continuous pathways between screw terminals — two pathways for each lever position. If not, then switch is faulty and must be replaced. (The arrangement of the pathways may differ, depending on the switch manufacturer. The photo above shows the three possible pathway arrangements.)

How to Test a Pilot-light Switch

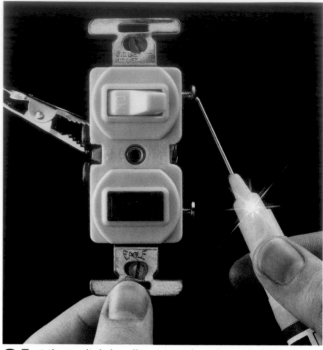

1 Test pilot light by flipping the switch lever to the ON position. Check to see if the light fixture or appliance is working. If the pilot light does not glow even though the switch operates the light fixture or appliance, then the pilot light is defective and the unit must be replaced.

2 Test the switch by disconnecting the unit. With the switch lever in the ON position, attach the tester clip to the top screw terminal on one side of the switch. Touch tester probe to top screw terminal on opposite side of the switch. If switch is good, tester will glow when switch is ON, but not when OFF.

How to Test a Timer Switch

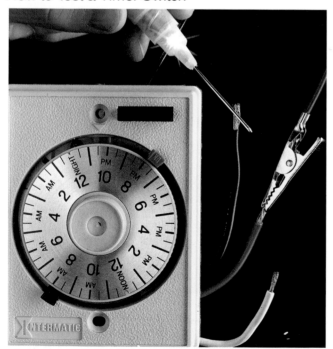

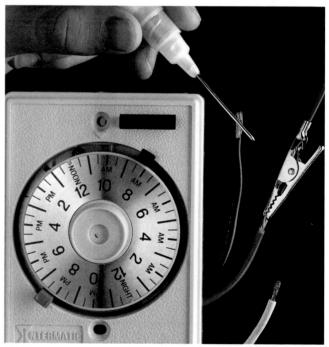

1 Attach the tester clip to the red wire lead on the timer switch, and touch the tester probe to the black hot lead. Rotate the timer dial clockwise until the ON tab passes the arrow marker. Tester should glow. If it does not, the switch is faulty and must be replaced.

2 Rotate the dial clockwise until the OFF tab passes the arrow marker. Tester should not glow. If it does, the switch is faulty and must be replaced.

How to Test Switch/receptacle

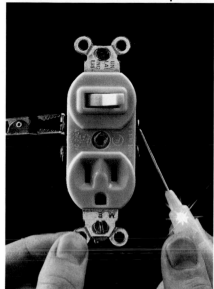

Attach tester clip to one of the top screw terminals. Touch the tester probe to the top screw terminal on the opposite side. Flip the switch lever from ON to OFF position. If the switch is working correctly, the tester will glow when the switch lever is ON, but not when OFF.

How to Test a Double Switch

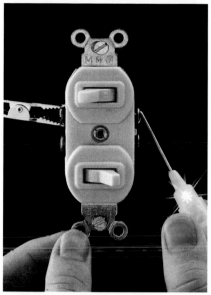

Test each half of switch by attaching the tester clip to one screw terminal, and touching the probe to the opposite side. Flip switch lever from ON to OFF position. If switch is good, tester glows when the switch lever is ON, but not when OFF. Repeat test with the remaining pair of screw terminals. If either half tests faulty, replace the unit.

How to Test a Time-delay Switch

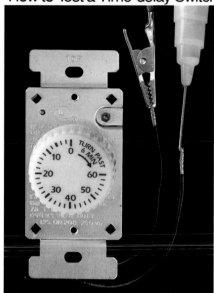

Attach tester clip to one of the wire leads, and touch the tester probe to the other lead. Set the timer for a few minutes. If switch is working correctly, the tester will glow until the time expires.

How to Test Manual Operation of Electronic Switches

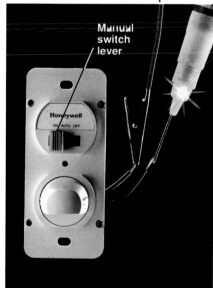

Automatic switch: Attach the tester clip to a black wire lead, and touch the tester probe to the other black lead. Flip the manual switch lever from ON to OFF position. If switch is working correctly, tester will glow when the switch lever is ON, but not when OFF.

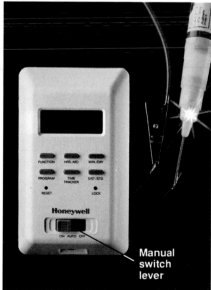

Programmable switch: Attach the tester clip to a wire lead, and touch the tester probe to the other lead. Flip the manual switch lever from ON to OFF position. If the switch is working correctly, the tester will glow when the switch lever is ON, but not when OFF.

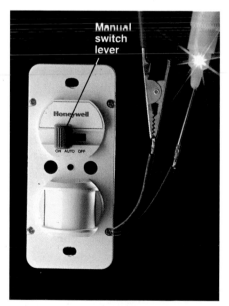

Motion-sensor switch: Attach the tester clip to a wire lead, and touch the tester probe to the other lead. Flip the manual switch lever from ON to OFF position. If the switch is working correctly, the tester will glow when the switch lever is ON, but not when OFF.

Fixing & Replacing Wall Switches

Most switch problems are caused by loose wire connections. If a fuse blows or a circuit breaker trips when a switch is turned on, a loose wire may be touching the metal box. Loose wires also can cause switches to overheat or buzz.

Switches sometimes fail because internal parts wear out. To check for wear, the switch must be removed entirely and tested for continuity (pages 42 to 45). If the continuity test shows the switch is faulty, replace it.

Everything You Need:

Tools: screwdriver, neon circuit tester, continuity tester, combination tool.

Materials: fine sandpaper, anti-oxidant paste (for aluminum wiring).

How to Fix or Replace a Single-pole Wall Switch

1 Turn off the power to the switch at the main service panel, then remove the switch coverplate.

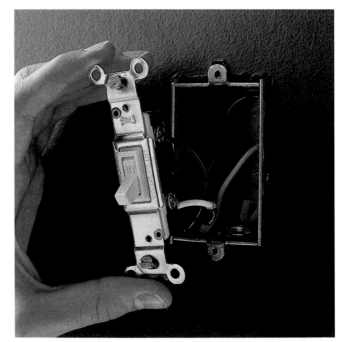

2 Remove the mounting screws holding the switch to the electrical box. Holding the mounting straps carefully, pull the switch from the box. Be careful not to touch any bare wires or screw terminals until the switch has been tested for power.

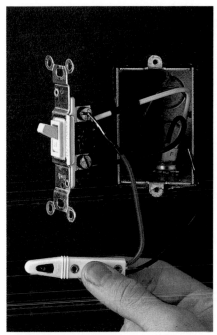

3 Test for power by touching one probe of the neon circuit tester to the grounded metal box or to the bare copper grounding wire, and touching other probe to each screw terminal. Tester should not glow. If it does, there is still power entering the box. Return to service panel and turn off correct circuit.

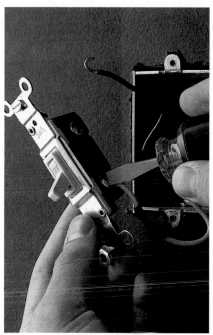

4 Disconnect the circuit wires and remove the switch. Test the switch for continuity (page 42), and buy a replacement if the switch is faulty. If circuit wires are too short, lengthen them by adding pigtail wires.

5 If wires are broken or nicked, clip off damaged portion, using a combination tool. Strip wires so there is about ¾" of bare wire at the end of each wire.

6 Clean the bare copper wires with fine sandpaper if they appear darkened or dirty. If wires are aluminum, apply an anti-oxidant paste before connecting the wires.

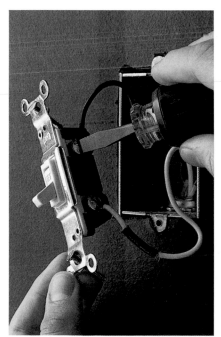

7 Connect the wires to the screw terminals on the switch. Tighten the screws firmly, but do not over-tighten. Overtightening may strip the screw threads

8 Remount the switch, carefully tucking the wires inside the box. Reattach the switch cover-plate and turn on the power to the switch at the main service panel.

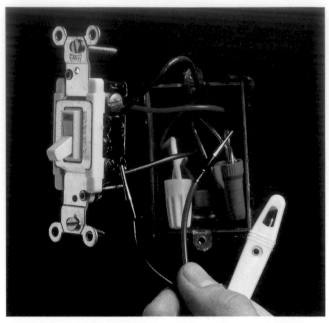

1 Turn off power to the switch at the main service panel, then remove the switch coverplate and mounting screws. Holding the mounting strap carefully, pull the switch from the box. Be careful not to touch the bare wires or screw terminals until they have been tested for power.

2 Test for power by touching one probe of the neon circuit tester to the grounded metal box or to the bare copper grounding wire, and touching the other probe to each screw terminal. Tester should not glow. If it does, there is still power entering the box. Return to the service panel and turn off the correct circuit.

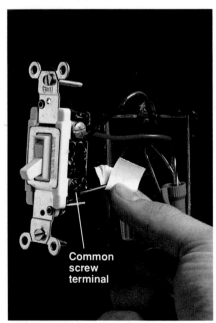

Common screw terminal

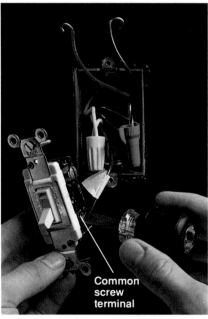

Common screw terminal

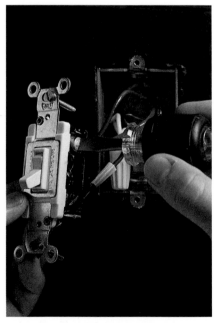

3 Locate dark common screw terminal, and use masking tape to label the "common" wire attached to it. Disconnect wires and remove switch. Test switch for continuity (page 43). If it tests faulty, buy a replacement. Inspect wires for nicks and scratches. If necessary, clip damaged wires and strip them (page 19).

4 Connect the common wire to the dark common screw terminal on the switch. On most three-way switches the common screw terminal is copper. Or, it may be labeled with the word COMMON stamped on the back of the switch.

5 Connect the remaining wires to the brass or silver screw terminals. These wires are interchangeable, and can be connected to either screw terminal. Carefully tuck the wires into the box. Remount the switch, and attach the coverplate. Turn on the power at the main service panel.

How to Fix or Replace a Four-way Wall Switch

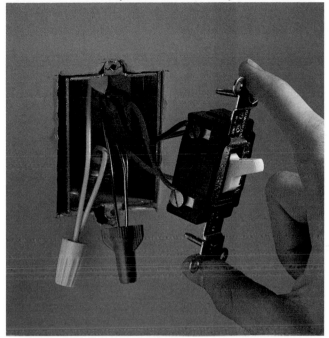

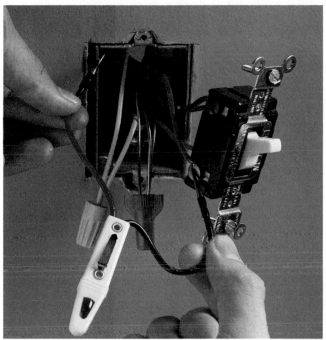

1 Turn off the power to the switch at the main service panel, then remove the switch coverplate and mounting screws. Holding the mounting strap carefully, pull the switch from the box. Be careful not to touch any bare wires or screw terminals until they have been tested for power.

2 Test for power by touching one probe of the circuit tester to the grounded metal box or bare copper grounding wire, and touching the other probe to each of the screw terminals. Tester should not glow. If it does, there is still power entering the box. Return to the service panel and turn off the correct circuit.

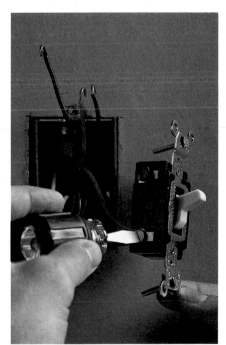

3 Disconnect the wires and inspect them for nicks and scratches. If necessary, clip damaged wires and strip them (page 19). Test the switch for continuity (page 43). Buy a replacement if the switch tests faulty.

4 Connect two wires of the same color to the brass screw terminals. On the switch shown above, the brass screw terminals are labeled LINE 1.

5 Attach remaining wires to copper screw terminals, marked LINE 2 on some switches. Carefully tuck the wires inside the switch box, then remount the switch and attach the coverplate. Turn on the power at the main service panel.

Toggle-type dimmer resembles standard switches. Toggle dimmers are available in both single-pole and three-way designs.

Dial-type dimmer is the most common style. Rotating the dial changes the light intensity.

Slide-action dimmer has an illuminated face that makes the switch easy to locate in the dark.

Automatic dimmer has an electronic sensor that adjusts the light fixture to compensate for the changing levels of natural light. An automatic dimmer also can be operated manually.

Dimmer Switches

A dimmer switch makes it possible to vary the brightness of a light fixture. Dimmers are often installed in dining rooms, recreation areas, or bedrooms.

Any standard single-pole switch can be replaced with a dimmer, as long as the switch box is of adequate size. Dimmer switches have larger bodies than standard switches. They also generate a small amount of heat that must dissipate. For these reasons, dimmers should not be installed in undersized electrical boxes, or in boxes that are crowded with circuit wires. Always follow the manufacturer's specifications for installation.

In lighting configurations that use three-way switches (pages 38 to 39), one of the three-way switches can be replaced with a special three-way dimmer. In this arrangement, all switches will turn the light fixture on and off, but light intensity will be controlled only from the dimmer switch.

Dimmer switches are available in several styles (photo, left). All types have wire leads instead of screw terminals, and they are connected to circuit wires using wire nuts. A few types have a green grounding lead that should be connected to the grounded metal box or to the bare copper grounding wires.

Everything You Need:

Tools: screwdriver, neon circuit tester, needlenose pliers.

Materials: wire nuts, masking tape.

How to Install a Dimmer Switch

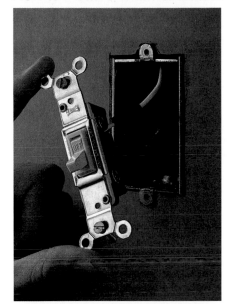

1 Turn off power to switch at the main service panel, then remove the coverplate and mounting screws. Holding the mounting straps carefully, pull switch from the box. Be careful not to touch bare wires or screw terminals until they have been tested for power.

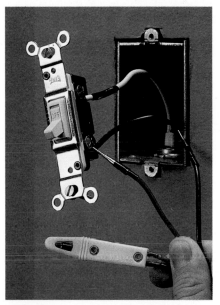

2 Test for power by touching one probe of neon circuit tester to the grounded metal box or to the bare copper grounding wires, and touching other probe to each screw terminal. Tester should not glow. If it does, there is still power entering the box. Return to the service panel and turn off the correct circuit.

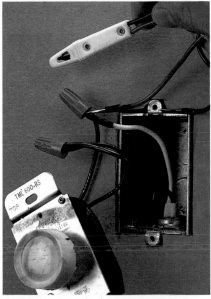

If replacing an old dimmer, test for power by touching one probe of circuit tester to the grounded metal box or bare copper grounding wires, and inserting the other probe into each wire nut. Tester should not glow. If it does, there is still power entering the box. Return to the service panel and turn off the correct circuit.

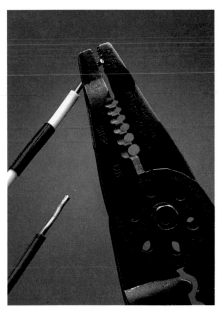

3 Disconnect the circuit wires and remove the switch. Straighten the circuit wires, and clip the ends, leaving about ½" of the bare wire end exposed.

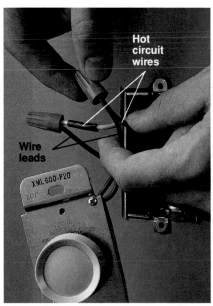

4 Connect the wire leads on the dimmer switch to the circuit wires, using wire nuts. The switch leads are interchangeable, and can be attached to either of the two circuit wires.

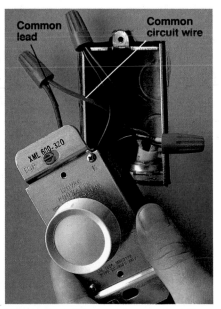

Three-way dimmer has an additional wire lead. This "common" lead is connected to the common circuit wire. When replacing a standard three-way switch with a dimmer, the common circuit wire is attached to the darkest screw terminal on the old switch (page 48).

The earliest receptacles were modifications of the screw-in type light bulb. This receptacle was used in the early 1900s.

Common Receptacle Problems

Household receptacles, also called outlets, have no moving parts to wear out, and usually last for many years without servicing. Most problems associated with receptacles are actually caused by faulty lamps and appliances, or their plugs and cords. However, the constant plugging in and removal of appliance cords can wear out the metal contacts inside a receptacle. Any receptacle that does not hold plugs firmly should be replaced.

A loose wire connection is another possible problem. A loose connection can spark (called arcing), trip a circuit breaker, or cause heat to build up in the receptacle box, creating a potential fire hazard.

Wires can come loose for a number of reasons. Everyday vibrations caused by walking across floors, or from nearby street traffic, may cause a connection to shake loose. In addition, because wires heat and cool with normal use, the ends of the wires will expand and contract slightly. This movement also may cause the wires to come loose from the screw terminal connections.

The polarized receptacle became standard in the 1920s. The different sized slots direct current flow for safety.

The ground-fault circuit-interrupter, or GFCI receptacle, is a modern safety device. When it detects slight changes in current, it instantly shuts off power.

Problem	Repair
Circuit breaker trips repeatedly, or fuse burns out immediately after being replaced.	1. Repair or replace worn or damaged lamp or appliance cord. 2. Move lamps or appliances to other circuits to prevent overloads (page 26). 3. Tighten any loose wire connections (pages 60 to 61). 4. Clean dirty or oxidized wire ends (page 60).
Lamp or appliance does not work.	1. Make sure lamp or appliance is plugged in. 2. Replace burned-out bulbs. 3. Repair or replace worn or damaged lamp or appliance cord. 4. Tighten any loose wire connections (pages 60 to 61). 5. Clean dirty or oxidized wire ends (page 60). 6. Repair or replace any faulty receptacle (pages 60 to 61).
Receptacle does not hold plugs firmly.	1. Repair or replace worn or damaged plugs (pages 77 to 78). 2. Replace faulty receptacle (pages 60 to 61).
Receptacle is warm to the touch, buzzes, or sparks when plugs are inserted or removed.	1. Move lamps or appliances to other circuits to prevent overloads (page 26). 2. Tighten any loose wire connections (pages 60 to 61). 3. Clean dirty or oxidized wire ends (page 60). 4. Replace faulty receptacle (pages 60 to 61).

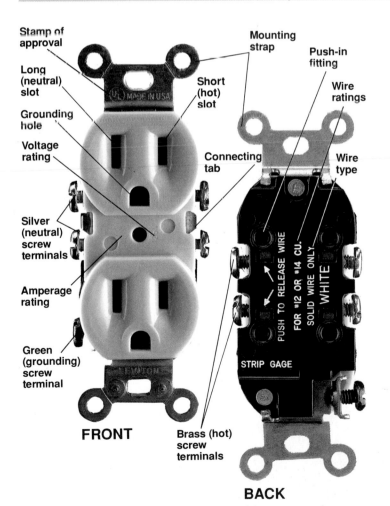

FRONT

Stamp of approval
Long (neutral) slot
Grounding hole
Voltage rating
Silver (neutral) screw terminals
Amperage rating
Green (grounding) screw terminal
Mounting strap
Short (hot) slot
Connecting tab
Brass (hot) screw terminals
Push-in fitting
Wire ratings
Wire type

BACK

The standard duplex receptacle has two halves for receiving plugs. Each half has a long (neutral) slot, a short (hot) slot, and a U-shaped grounding hole. The slots fit the wide prong, narrow prong, and grounding prong of a three-prong plug. This ensures that the connection between receptacle and plug will be polarized and grounded for safety (page 12).

Wires are attached to the receptacle at screw terminals or push-in fittings. A connecting tab between the screw terminals allows a variety of different wiring configurations. Receptacles also include mounting straps for attaching to electrical boxes.

Stamps of approval from testing agencies are found on the front and back of the receptacle. Look for the symbol UL or UND. LAB. INC. LIST to make sure the receptacle meets the strict standards of Underwriters Laboratories.

The receptacle is marked with ratings for maximum volts and amps. The common receptacle is marked 15A, 125V. Receptacles marked CU or COPPER are used with solid copper wire. Those marked CU-CLAD ONLY are used with copper-coated aluminum wire. Only receptacles marked CO/ALR may be used with solid aluminum wiring (page 18). Receptacles marked AL/CU no longer may be used with aluminum wire according to code.

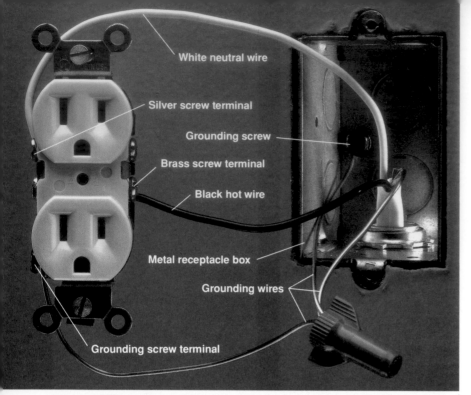

White neutral wire

Silver screw terminal

Grounding screw

Brass screw terminal

Black hot wire

Metal receptacle box

Grounding wires

Grounding screw terminal

Single cable entering the box indicates end-of-run wiring. The black hot wire is attached to a brass screw terminal, and the white neutral wire is connected to a silver screw terminal. If the box is metal, the grounding wire is pigtailed to the grounding screws of the receptacle and the box. In a plastic box, the grounding wire is attached directly to the grounding screw terminal of the receptacle.

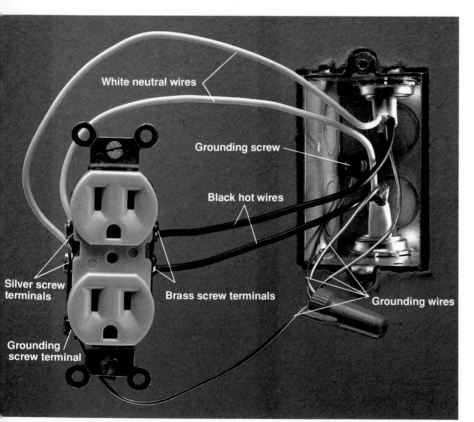

White neutral wires

Grounding screw

Black hot wires

Silver screw terminals

Brass screw terminals

Grounding wires

Grounding screw terminal

Two cables entering the box indicate middle-of-run wiring. Black hot wires are connected to brass screw terminals, and white neutral wires to silver screw terminals. The grounding wire is pigtailed to the grounding screws of the receptacle and the box.

Receptacle Wiring

A 125-volt duplex receptacle can be wired to the electrical system in a number of ways. The most common are shown on these pages.

Wiring configurations may vary slightly from these photographs, depending on the kind of receptacle used, the type of cable, or the technique of the electrician who installed the wiring. To make dependable repairs or replacements, use masking tape and label each wire according to its location on the terminals of the existing receptacle.

Receptacles are wired as either **end-of-run** or **middle-of-run**. These two basic configurations are easily identified by counting the number of cables entering the receptacle box. End-of-run wiring has only one cable, indicating that the circuit ends. Middle-of-run wiring has two cables, indicating that the circuit continues on to other receptacles, switches, or fixtures.

A split-circuit receptacle is shown on the opposite page. Each half of a split-circuit receptacle is wired to a separate circuit. This allows two appliances of high wattage to be plugged into the same receptacle without blowing a fuse or tripping a breaker. This wiring configuration is similar to a receptacle that is controlled by a wall switch. Code requires a **switch-controlled receptacle** in any room that does not have a built-in light fixture operated by a wall switch.

Split-circuit and switch-controlled receptacles are connected to two hot wires, so use caution during repairs or replacements. Make sure the connecting tab between the hot screw terminals is removed.

Two-slot receptacles are common in older homes. There is no grounding wire attached to the receptacle, but the box may be grounded with armored cable or conduit (page 16).

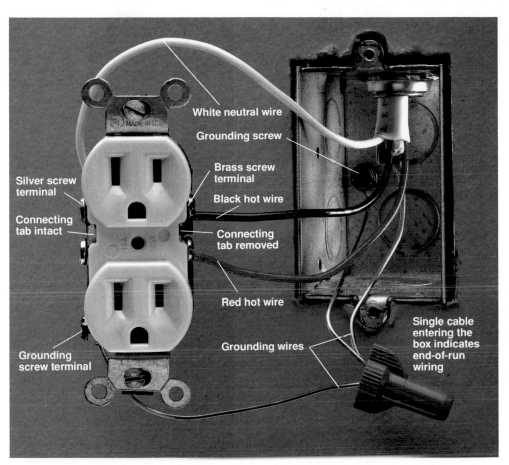

Split-circuit receptacle is attached to a black hot wire, a red hot wire, a white neutral wire, and a bare grounding wire. The wiring is similar to a switch-controlled receptacle.

The hot wires are attached to the brass screw terminals, and the connecting tab or fin between the brass terminals is removed. The white wire is attached to a silver screw terminal, and the connecting tab on the neutral side remains intact. The grounding wire is pigtailed to the grounding screw terminal of the receptacle, and to the grounding screw attached to the box.

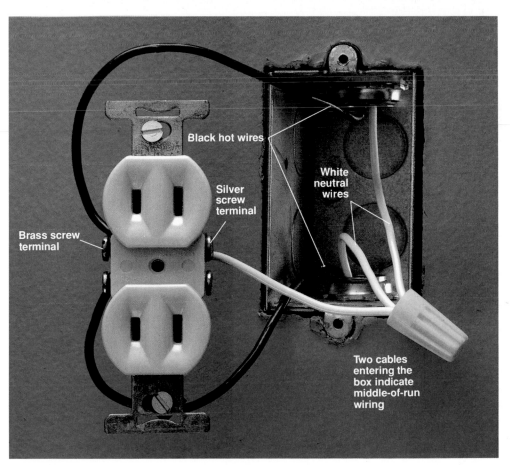

Two-slot receptacle is often found in older homes. The black hot wires are connected to the brass screw terminals, and the white neutral wires are pigtailed to a silver screw terminal.

Two-slot receptacles may be replaced with three-slot types, but only if a means of grounding exists at the receptacle box.

Basic Types of Receptacles

Several different types of receptacles are found in the typical home. Each has a unique arrangement of slots that accepts only a certain kind of plug, and each is designed for a specific job.

Household receptacles provide two types of voltage: normal and high voltage. Although voltage ratings have changed slightly over the years, normal receptacles should be rated for 110, 115, 120, or 125 volts. For purposes of replacement, these ratings are considered identical. High-voltage receptacles are rated at 220, 240, or 250 volts. These ratings are considered identical.

When replacing a receptacle, check the amperage rating of the circuit at the main service panel, and buy a receptacle with the correct amperage rating (page 24).

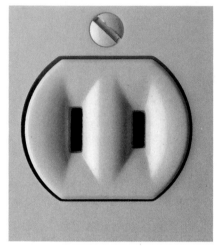

15 amps, 125 volts. Polarized two-slot receptacle is common in homes built before 1960. Slots are different sizes to accept polarized plugs.

15 amps, 125 volts. Three-slot grounded receptacle has two different size slots and a U-shaped hole for grounding. It is required in all new wiring installations.

20 amps, 125 volts. This three-slot grounded receptacle features a special T-shaped slot. It is installed for use with large appliances or portable tools that require 20 amps of current.

15 amps, 250 volts. This receptacle is used primarily for window air conditioners. It is available as a single unit, or as half of a duplex receptacle, with the other half wired for 125 volts.

30 amps, 125/250 volts. This receptacle is used for clothes dryers. It provides high-voltage current for heating coils, and 125-volt current to run lights and timers.

50 amps, 125/250 volts. This receptacle is used for ranges. The high-voltage current powers heating coils, and the 125-volt current runs clocks and lights.

Older Receptacles

Older receptacles may look different from more modern types, but most will stay in good working order. Follow these simple guidelines for evaluating or replacing older receptacles:

• Never replace an older receptacle with one of a different voltage or higher amperage rating.
• Any two-slot, unpolarized receptacle should be replaced with a two- or three-slot polarized receptacle.
• If no means of grounding is available at the receptacle box, install a GFCI (pages 62 to 63).
• If in doubt, seek the advice of a qualifified electrician.

Never alter the prongs of a plug to fit an older receptacle. Altering the prongs may remove the grounding or polarizing features of the plug.

Unpolarized receptacles have slots that are the same length. Modern plug types may not fit these receptacles. Never modify the prongs of a polarized plug to fit the slots of an unpolarized receptacle.

Surface-mounted receptacles were popular in the 1940s and 1950s for their ease of installation. Wiring often ran in the back of hollowed-out base moldings. Surface-mounted receptacles are usually ungrounded.

Ceramic duplex receptacles were manufactured in the 1930s. They are polarized but ungrounded, and they can be wired for either 125 volts or 250 volts.

Twist-lock receptacles are designed to be used with plugs that are inserted and rotated. A small tab on the end of one of the prongs prevents the plug from being pulled from the receptacle.

Ceramic duplex receptacle has a unique hourglass shape. The receptacle shown above is rated for 250 volts but only 5 amps, and would not be allowed by today's electrical codes.

Metal probes

Insulated handles

Bulb

Testing Receptacles for Power, Grounding & Polarity

Test for power to make sure that live voltage is not reaching the receptacle during a repair or replacement project.

Test for grounding to plan receptacle replacements. The test for grounding will indicate how an existing receptacle is wired, and whether a replacement receptacle should be a two-slot polarized receptacle, a grounded three-slot receptacle, or a GFCI.

If the test indicates that the hot and neutral wires are reversed, make sure the wires are installed correctly on the replacement receptacle.

Test for hot wires if you need to confirm which wire is carrying live voltage.

An inexpensive neon circuit tester makes it easy to perform these tests. It has a small bulb that glows when electrical power flows through it.

Remember that the tester only glows when it is part of a complete circuit. For example, if you touch one probe to a hot wire and do not touch anything with the other probe, the tester will not glow, even though the hot wire is carrying power. When using the tester, take care not to touch the metal probes.

When testing for power or grounding, always confirm any negative (tester does not glow) results by removing the coverplate and examining the receptacle to make sure all wires are intact and properly connected. Do not touch any wires without first turning off the power at the main service panel.

Everything You Need:

Tools: neon circuit tester, screwdriver.

How to Test a Receptacle for Power

1 Turn off power at the main service panel. Place one probe of the tester in each slot of the receptacle. The tester should not glow. If it does glow, the correct circuit has not been turned off at the main service panel. Test both ends of a duplex receptacle. Remember that this is a preliminary test. You must confirm that power is off by removing the coverplate and testing for power at the receptacle wires (step 2).

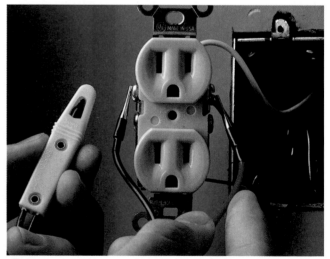

2 Remove the receptacle coverplate. Loosen the mounting screws and carefully pull the receptacle from its box. Take care not to touch any wires. Touch one probe of the neon tester to a brass screw terminal, and one probe to a silver screw terminal. The tester should not glow. If it does, you must shut off the correct circuit at the service panel. If wires are connected to both sets of terminals, test both sets.

This photo shows hot and neutral wires that are reversed.

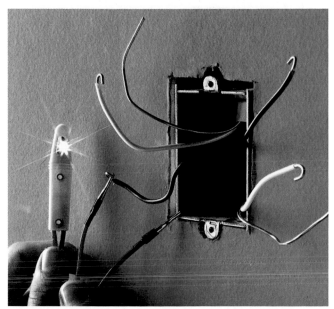

Test a three-slot receptacle for grounding. With the power on, place one probe of the tester in the short (hot) slot, and the other in the U-shaped grounding hole. The tester should glow. If it does not glow, place a probe in the long (neutral) slot and one in the grounding hole. If the tester glows, the hot and neutral wires are reversed (page 121). If tester does not glow in either position, the receptacle is not grounded.

Test for hot wires. Occasionally, you may need to determine which wire is hot. With the power turned off, carefully separate all ends of wires so that they do not touch each other or anything else. Restore power to the circuit at the main service panel. Touch one probe of the neon tester to the bare grounding wire or grounded metal box, and the other probe to the ends of each of the wires. Check all wires. If the tester glows, the wire is hot. Label the hot wire for identification, and turn off power at the service panel before continuing work.

How to Test a Two-slot Receptacle for Grounding

1 With the power turned on, place one probe of the neon tester in each slot. The tester should glow. If it does not glow, then there is no power to the receptacle.

2 Place one probe of the tester in the short (hot) slot, and touch the other probe to the coverplate screw. The screw head must be free of paint, dirt, and grease. If the tester glows, the receptacle box is grounded. If it does not glow, proceed to step 3.

3 Place one probe of the tester in the long (neutral) slot, and touch the other to the coverplate screw. If the tester glows, the receptacle box is grounded but hot and neutral wires are reversed (page 121). If tester does not glow, the box is not grounded.

Repairing & Replacing Receptacles

Receptacles are easy to repair. After shutting off power to the receptacle circuit, remove the coverplate and inspect the receptacle for any obvious problems such as a loose or broken connection, or wire ends that are dirty or oxidized. Remember that a problem at one receptacle may affect other receptacles in the same circuit. If the cause of a faulty receptacle is not readily apparent, test other receptacles in the circuit for power (page 58).

When replacing a receptacle, check the amperage rating of the circuit at the main service panel, and buy a replacement receptacle with the correct amperage rating (page 26).

When installing a new receptacle, always test for grounding (pages 58 to 59). Never install a three-slot receptacle where no grounding exists. Instead, install a two-slot polarized, or GFCI receptacle.

Everything You Need:

Tools: neon circuit tester, screwdriver, vacuum cleaner (if needed).

Materials: fine sandpaper, anti-oxidant paste (if needed).

1 Turn off power at the main service panel. Test the receptacle for power with a neon circuit tester (page 58). Test both ends of a duplex receptacle. Remove the coverplate, using a screwdriver.

2 Remove the mounting screws that hold the receptacle to the box. Carefully pull the receptacle from the box. Take care not to touch any bare wires.

3 Confirm that the power to the receptacle is off (page 58), using a neon circuit tester. If wires are attached to both sets of screw terminals, test both sets. The tester should not glow. If it does, you must turn off the correct circuit at the service panel.

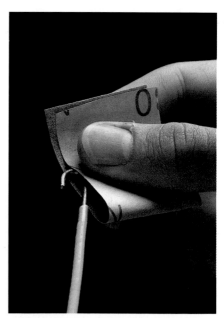

4 If the ends of the wires appear darkened or dirty, disconnect them one at a time, and clean them with fine sandpaper. If the wires are aluminum, apply an anti-oxidant paste before reconnecting. Anti-oxidant paste is available at hardware stores.

5 Tighten all connections, using a screwdriver. Take care not to overtighten and strip the screws.

6 Check the box for dirt or dust and, if necessary, clean it with a vacuum cleaner and narrow nozzle attachment.

7 Reinstall the receptacle, and turn on power at the main service panel. Test the receptacle for power with a neon circuit tester. If the receptacle does not work, check other receptacles in the circuit before making a replacement.

How to Replace a Receptacle

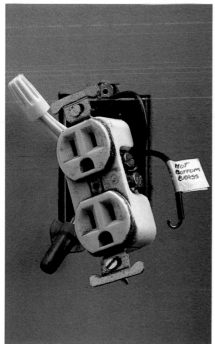

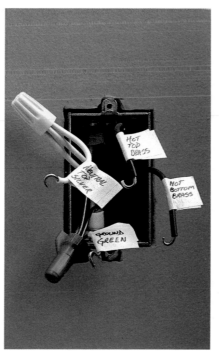

1 To replace a receptacle, repeat steps 1 to 3 on the opposite page. With the power off, label each wire for its location on the receptacle screw terminals, using masking tape and a felt-tipped pen.

2 Disconnect all wires and remove the receptacle.

3 Replace the receptacle with one rated for the correct amperage and voltage (page 24). Replace coverplate, and turn on power. Test receptacle with a neon circuit tester (pages 58 to 59).

GFCI Receptacles

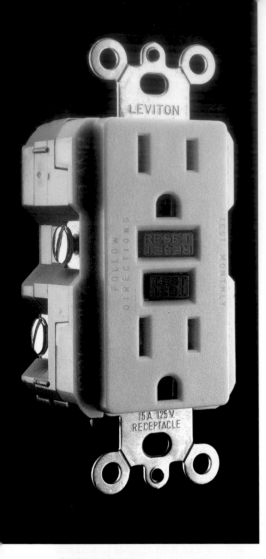

The ground-fault circuit-interrupter (GFCI) receptacle is a safety device. It protects against electrical shock caused by a faulty appliance, or a worn cord or plug. The GFCI senses small changes in current flow and can shut off power in as little as 1/40 of a second.

When you are updating wiring, installing new circuits, or replacing receptacles, GFCIs are now required in bathrooms, kitchens, garages, crawl spaces, unfinished basements, and outdoor receptacle locations. GFCIs are easily installed as safety replacements for any standard duplex receptacle. Consult your local codes for any requirements regarding the installation of GFCI receptacles.

The GFCI receptacle may be wired to protect only itself (single location), or it can be wired to protect all receptacles, switches, and light fixtures from the GFCI "forward" to the end of the circuit (multiple locations). The GFCI cannot protect those devices that exist between the GFCI location and the main service panel.

Because the GFCI is so sensitive, it is most effective when wired to protect a single location. The more receptacles any one GFCI protects, the more susceptible it is to "phantom tripping," shutting off power because of tiny, normal fluctuations in current flow.

Everything You Need:

Tools: neon circuit tester, screwdriver.

Materials: wire nuts, masking tape.

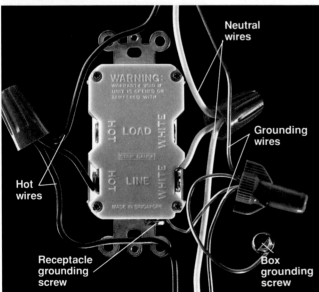

A GFCI wired for single-location protection (shown from the back) has hot and neutral wires connected only to the screw terminals marked LINE. A GFCI connected for single-location protection may be wired as either an end-of-run or middle-of-run configuration (page 54).

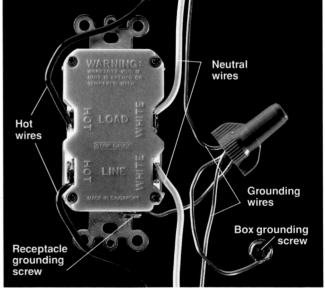

A GFCI wired for multiple-location protection (shown from the back) has one set of hot and neutral wires connected to the LINE pair of screw terminals, and the other set connected to the LOAD pair of screw terminals. A GFCI receptacle connected for multiple-location protection may be wired only as a middle-of-run configuration.

How to Install a GFCI for Single-location Protection

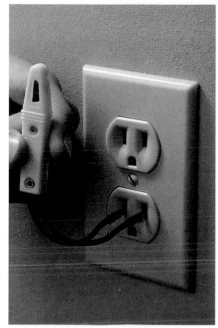

1 Shut off power to the receptacle at the main service panel. Test for power with a neon circuit tester (page 58). Be sure to check both halves of the receptacle.

2 Remove coverplate. Loosen mounting screws, and gently pull receptacle from the box. Do not touch wires. Confirm power is off with a circuit tester (page 58).

3 Disconnect all white neutral wires from the silver screw terminals of the old receptacle.

4 Pigtail all the white neutral wires together, and connect the pigtail to the terminal marked WHITE LINE on the GFCI (see photo on opposite page).

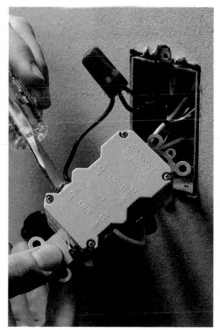

5 Disconnect all black hot wires from the brass screw terminals of the old receptacle. Pigtail these wires together and connect them to the terminal marked HOT LINE on the GFCI.

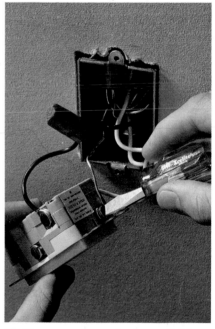

6 If a grounding wire is available, disconnect it from the old receptacle, and reconnect it to the green grounding screw terminal of the GFCI. Mount the GFCI in the receptacle box, and reattach the coverplate. Restore power and test the GFCI according to the manufacturer's instructions.

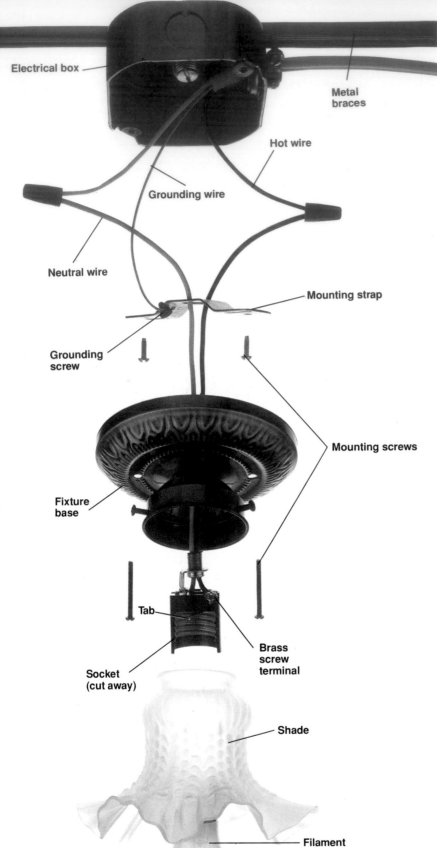

Electrical box

Metal braces

Hot wire

Grounding wire

Neutral wire

Mounting strap

Grounding screw

Mounting screws

Fixture base

Tab

Brass screw terminal

Socket (cut away)

Shade

Filament

Repairing & Replacing Incandescent Light Fixtures

In a typical incandescent light fixture, a black hot wire is connected to a brass screw terminal on the socket. Power flows to a small tab at the bottom of the metal socket and through a metal filament inside the bulb. The power heats the filament and causes it to glow. The current then flows through the threaded portion of the socket and through the white neutral wire back to the main service panel.

Incandescent light fixtures are attached permanently to ceilings or walls. They include wall-hung sconces, ceiling-hung globe fixtures, recessed light fixtures, and chandeliers. Most incandescent light fixtures are easy to repair, using basic tools and inexpensive parts.

Track lights, another type of incandescent light fixture, are difficult to fix and should be repaired or replaced by an electrician.

If a light fixture fails, always check the light bulb to make sure it is screwed in tightly and is not burned out. A faulty light bulb is the most common cause of light fixture failure. If the light fixture is controlled by a wall switch, also check the switch as a possible source of problems (pages 34 to 51).

Light fixtures can fail because the sockets or built-in switches wear out. Some fixtures have sockets and switches that can be removed for minor repairs. These parts are held to the base of the fixture with mounting screws or clips. Other fixtures have sockets and switches that are joined permanently to the base. If this type of fixture fails, purchase and install a new light fixture.

Damage to light fixtures often occurs because homeowners install light bulbs with wattage ratings that are too high. Prevent overheating and light fixture failures by using only light bulbs that match the wattage ratings printed on the fixtures.

Techniques for repairing fluorescent lights are different from those for incandescent lights. Refer to pages 72 to 76 to repair a fluorescent light fixture.

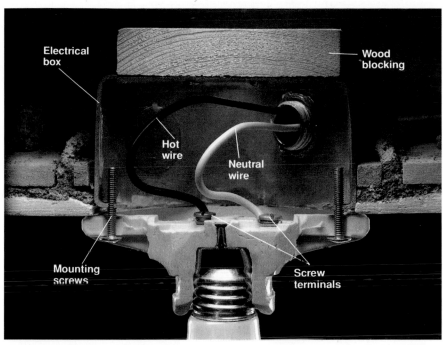

Everything You Need:

Tools: neon circuit tester, screwdrivers, continuity tester, combination tool.

Materials: replacement parts, as needed.

Before 1959, incandescent lighting fixtures (shown cut away) often were mounted directly to an electrical box or to plaster lath. Electrical codes now require that lighting fixtures be attached to mounting straps that are anchored to the electrical boxes (page opposite). If you have a light fixture attached to plaster lath, install an approved electrical box with a mounting strap to support the fixture (pages 30 to 33).

Problem	Repair
Wall- or ceiling-mounted fixture flickers, or does not light.	1. Check for faulty light bulb. 2. Check wall switch; repair or replace, if needed (pages 34 to 51). 3. Check for loose wire connections in electrical box. 4. Test socket, and replace, if needed (pages 66 to 67). 5. Replace light fixture (page 68).
Built-in switch on fixture does not work.	1. Check for faulty light bulb. 2. Check for loose wire connections on switch. 3. Replace switch (page 67). 4. Replace light fixture (page 68).
Chandelier flickers or does not light.	1. Check for faulty light bulb. 2. Check wall switch, and repair or replace, if needed (pages 34 to 51). 3. Check for loose wire connections in electrical box. 4. Test socket, and replace, if needed.
Recessed fixture flickers or does not light.	1. Check for faulty light bulb. 2. Check wall switch; repair or replace, if needed (pages 34 to 51). 3. Check for loose wire connections in electrical box. 4. Test socket, and repair or replace, if needed (pages 69 to 70). 5. Replace recessed light fixture (page 71).

How to Remove a Light Fixture & Test a Socket

1 Turn off the power to the light fixture at the main service panel. Remove the light bulb and any shade or globe, then remove the mounting screws holding the fixture base to the electrical box or mounting strap. Carefully pull the fixture base away from box.

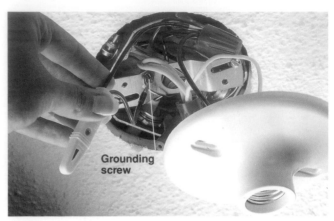

2 Test for power by touching one probe of a neon circuit tester to green grounding screw, then inserting other probe into each wire nut. Tester should not glow. If it does, there is still power entering box. Return to the service panel and turn off power to correct circuit.

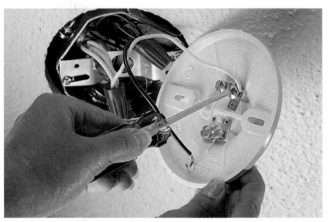

3 Disconnect the light fixture base by loosening the screw terminals. If fixture has wire leads instead of screw terminals, remove the light fixture base by unscrewing the wire nuts.

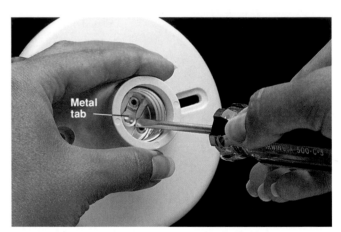

4 Adjust the metal tab at the bottom of the fixture socket by prying it up slightly with a small screwdriver. This adjustment will improve the contact between the socket and the light bulb.

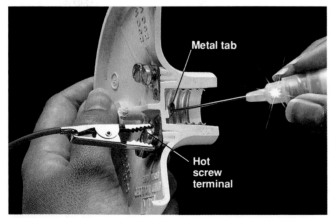

5 Test the socket (shown cut away) by attaching the clip of a continuity tester to the hot screw terminal (or black wire lead), and touching probe of tester to metal tab in bottom of socket. Tester should glow. If not, socket is faulty and must be replaced.

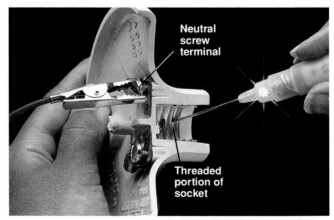

6 Attach tester clip to neutral screw terminal (or white wire lead), and touch probe to threaded portion of socket. Tester should glow. If not, socket is faulty and must be replaced. If socket is permanently attached, replace the fixture (page 68).

How to Replace a Socket

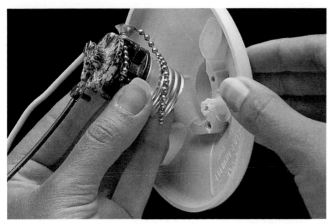

1 Remove light fixture (steps 1 to 3, page opposite). Remove the socket from the fixture. Socket may be held by a screw, clip, or retaining ring. Disconnect wires attached to the socket.

2 Purchase an identical replacement socket. Connect white wire to silver screw terminal on socket, and connect black wire to brass screw terminal. Attach socket to fixture base, and reinstall fixture.

How to Test & Replace a Built-in Light Switch

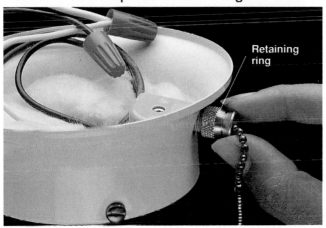

Retaining ring

1 Remove light fixture (steps 1 to 3, page opposite). Unscrew the retaining ring holding the switch.

Switch leads

2 Label the wires connected to the switch leads. Disconnect the switch leads and remove switch.

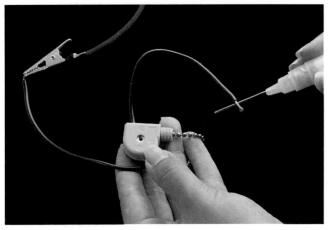

3 Test switch by attaching clip of continuity tester to one of the switch leads, and holding tester probe to the other lead. Operate switch control. If switch is good, tester will glow when switch is in one position, but not both.

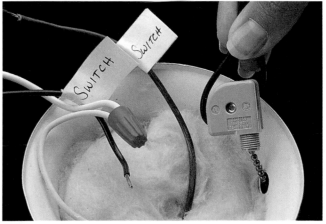

4 If the switch is faulty, purchase and install an exact duplicate switch. Remount the light fixture, and turn on the power at the main service panel.

How to Replace an Incandescent Light Fixture

1 Turn off the power and remove the old light fixture, following the directions for standard light fixtures (page 66, steps 1 to 3).

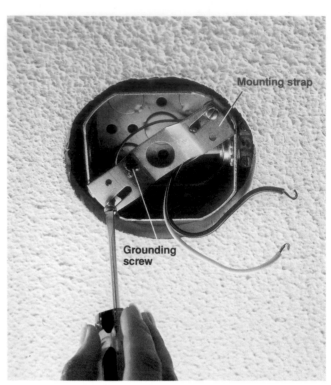

2 Attach a mounting strap to the electrical box, if box does not already have one. The mounting strap, included with the new light fixture, has a preinstalled grounding screw.

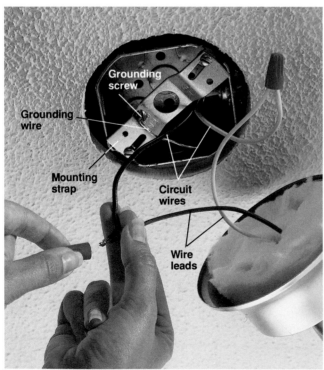

3 Connect the circuit wires to the base of new fixture, using wire nuts. Connect the white wire lead to the white circuit wire, and the black wire lead to the black circuit wire. Pigtail the bare copper grounding wire to the grounding screw on the mounting strap.

4 Attach the light fixture base to the mounting strap, using the mounting screws. Attach the globe, and install a light bulb with a wattage rating that is the same or lower than the rating indicated on the fixture. Turn on the power at the main service panel.

Repairing & Replacing Recessed Light Fixtures

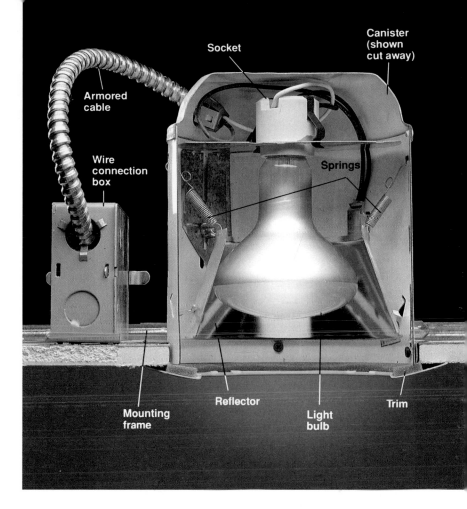

Most problems with recessed light fixtures occur because heat builds up inside the metal canister and melts the insulation on the socket wires. On some recessed light fixtures, sockets with damaged wires can be removed and replaced. However, most newer fixtures have sockets that cannot be removed. With this type, you will need to buy a new fixture if the socket wires are damaged.

When buying a new recessed light fixture, choose a replacement that matches the old fixture. Install the new fixture in the metal mounting frame that is already in place.

Make sure building insulation is at least 3" away from the metal canister on a recessed light fixture. Insulation that fits too closely traps heat and can damage the socket wires.

How to Remove & Test a Recessed Light Fixture

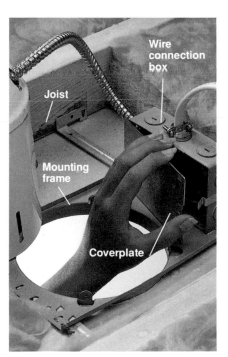

1 Turn off the power to the light fixture at the main service panel. Remove the trim, light bulb, and reflector. The reflector is held to the canister with small springs or mounting clips.

2 Loosen the screws or clips holding the canister to the mounting frame. Carefully raise the canister and set it away from the frame opening.

3 Remove the coverplate on the wire connection box. The box is attached to the mounting frame between the ceiling joists.

(continued next page)

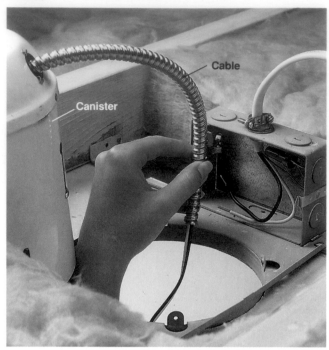

4 Test for power by touching one probe of neon circuit tester to grounded wire connection box, and inserting other probe into each wire nut. Tester should not glow. If it does, there is still power entering box. Return to the service panel and turn off correct circuit.

5 Disconnect the white and black circuit wires by removing the wire nuts. Pull the armored cable from the wire connection box. Remove the canister through the frame opening.

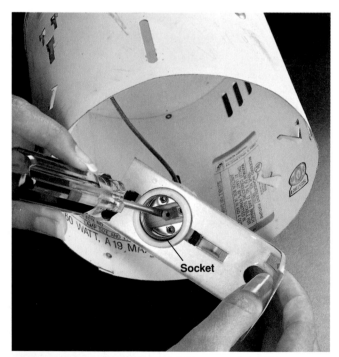

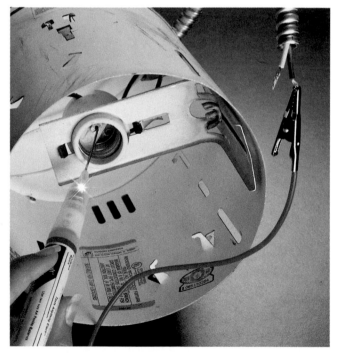

6 Adjust the metal tab at the bottom of the fixture socket by prying it up slightly with a small screwdriver. This adjustment will improve contact with the light bulb.

7 Test the socket by attaching clip of a continuity tester to the black fixture wire and touching tester probe to the metal tab in bottom of the socket. Attach tester clip to white fixture wire, and touch probe to threaded metal socket. Tester should glow for both tests. If not, then socket is faulty. Replace the socket (page 67), or install a new light fixture (page opposite).

How to Replace a Recessed Light Fixture

1 Remove the old light fixture (pages 69 to 70). Buy a new fixture that matches the old fixture. Although new light fixture comes with its own mounting frame, it is easier to mount the new fixture using the frame that is already in place.

2 Set the fixture canister inside the ceiling cavity, and thread the fixture wires through the opening in the wire connection box. Push the armored cable into the wire connection box to secure it.

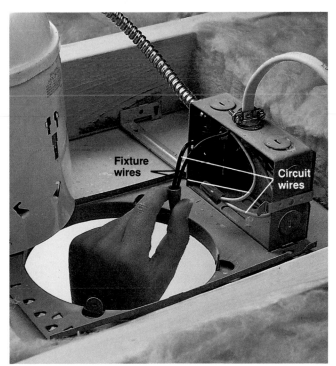

Fixture wires

Circuit wires

3 Connect the white fixture wire to the white circuit wire, and the black fixture wire to the black circuit wire, using wire nuts. Attach the coverplate to the wire connection box. Make sure any building insulation is at least 3" from canister and wire connection box.

4 Position the canister inside the mounting frame, and attach the mounting screws or clips. Attach the reflector and trim. Install a light bulb with a wattage rating that is the same or lower than rating indicated on the fixture. Turn on power at main service panel.

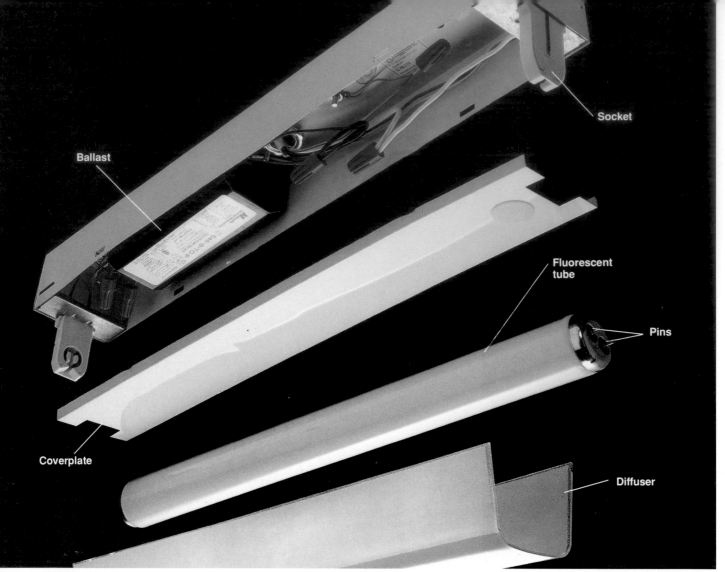

Ballast

Socket

Coverplate

Fluorescent tube

Pins

Diffuser

A fluorescent light works by directing electrical current through a special gas-filled **tube** that glows when energized. A white transluscent **diffuser** protects the fluorescent tube and softens the light. A **coverplate** protects a special transformer, called a **ballast**. The ballast regulates the flow of 120-volt household current to the **sockets**. The sockets transfer power to metal **pins** that extend into the tube.

Repairing & Replacing Fluorescent Lights

Fluorescent lights are relatively trouble-free, and use less energy than incandescent lights. A typical fluorescent tube lasts about three years, and produces two to four times as much light per watt as a standard incandescent light bulb.

The most frequent problem with a fluorescent light fixture is a worn-out tube. If a fluorescent light fixture begins to flicker, or does not light fully, remove and examine the tube. If the tube has bent or broken pins, or black discoloration near the ends, replace it. Light gray discoloration is normal in working fluorescent tubes. When replacing an old tube, read the wattage rating printed on the glass surface, and buy a new tube with a matching rating.

Never dispose of old tubes by breaking them. Fluorescent tubes contain a small amount of hazardous mercury. Check with your local environmental control agency or health department for disposal guidelines.

Fluorescent light fixtures also can malfunction if the sockets are cracked or worn. Inexpensive replacement sockets are available at any hardware store, and can be installed in a few minutes.

If a fixture does not work even after the tube and sockets have been serviced, the ballast probably is defective. Faulty ballasts may leak a black, oily substance, and can cause a fluorescent light

Problem	Repair
Tube flickers, or lights partially.	1. Rotate tube to make sure it is seated properly in the sockets. 2. Replace tube (page 74) and starter (where present) if tube is discolored or if pins are bent or broken. 3. Replace ballast (page 76) if replacement cost is reasonable. Otherwise, replace the entire fixture.
Tube does not light.	1. Check wall switch; repair or replace, if needed (pages 34 to 51). 2. Rotate tube to make sure it is seated properly in the sockets. 3. Replace tube (page 74) and starter (where present) if tube is discolored or if pins are bent or broken. 4. Replace sockets if they are chipped, or if tube does not seat properly (page 75). 5. Replace the ballast (page 76) or the entire fixture.
Noticeable black substance around ballast.	Replace ballast (page 76) if replacement cost is reasonable. Otherwise, replace the entire fixture.
Fixture hums.	Replace ballast (page 76) if replacement cost is reasonable. Otherwise, replace the entire fixture.

fixture to make a loud humming sound. Although ballasts can be replaced, always check prices before buying a new ballast. It may be cheaper to purchase and install a new fluorescent fixture rather than to replace the ballast in an old fluorescent light fixture.

Everything You Need:

Tools: screwdriver, ratchet wrench, combination tool, neon circuit tester.

Materials: replacement tubes, starters, or ballast (if needed); replacement fluorescent light fixture (if needed).

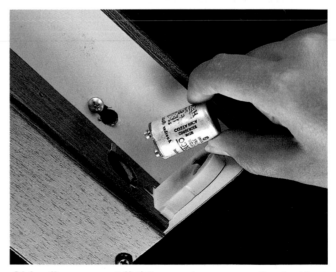

Older fluorescent lights may have a small cylindrical device, called a starter, located near one of the sockets. When a tube begins to flicker, replace both the tube and the starter. Turn off the power, then remove the starter by pushing it slightly and turning counterclockwise. Install a replacement that matches the old starter.

How to Replace a Fluorescent Tube

1 Turn off power to the light fixture at the main service panel. Remove the diffuser to expose the fluorescent tube.

2 Remove the fluorescent tube by rotating it ¼ turn in either direction and sliding the tube out of the sockets. Inspect the pins at the end of the tube. Tubes with bent or broken pins should be replaced.

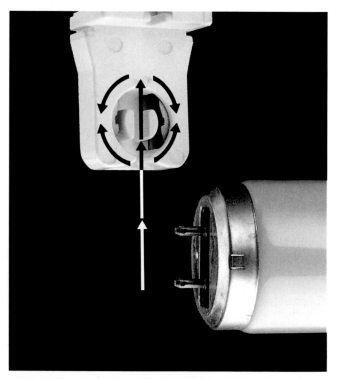

3 Inspect the ends of the fluorescent tube for discoloration. New tube in good working order (top) shows no discoloration. Normal, working tube (middle) may have gray color. A worn-out tube (bottom) shows black discoloration.

4 Install a new tube with the same wattage rating as the old tube. Insert the tube so that pins slide fully into sockets, then twist tube ¼ turn in either direction until it is locked securely. Reattach the diffuser, and turn on the power at the main service panel.

How to Replace a Socket

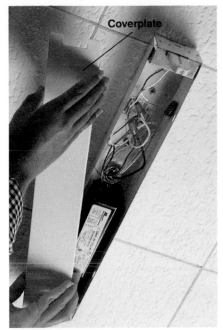

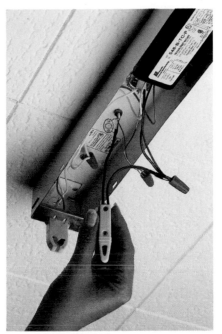

1 Turn off the power at the main service panel. Remove the diffuser, fluorescent tube, and the coverplate.

2 Test for power by touching one probe of a neon circuit tester to the grounding screw, and inserting the other probe into each wire nut. Tester should not glow. If it does, power is still entering the box. Return to the service panel and turn off correct circuit.

3 Remove the faulty socket from the fixture housing. Some sockets slide out, while others must be unscrewed.

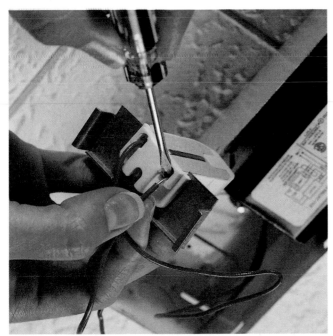

4 Disconnect wires attached to socket. For push-in fittings (above) remove the wires by inserting a small screwdriver into the release openings. Some sockets have screw terminal connections, while others have preattached wires that must be cut before the socket can be removed.

5 Purchase and install a new socket. If socket has preattached wire leads, connect the leads to the ballast wires using wire nuts. Replace coverplate and diffuser, then turn on power at the main service panel.

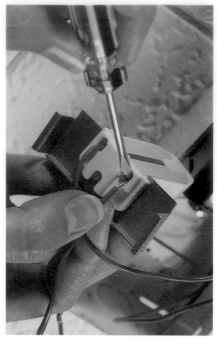

1 Turn off the power at the main service panel, then remove the diffuser, fluorescent tube, and coverplate. Test for power, using a neon circuit tester (step 2, page 75).

2 Remove the sockets from the fixture housing by sliding them out, or by removing the mounting screws and lifting the sockets out.

3 Disconnect the wires attached to the sockets by pushing a small screwdriver into the release openings (above), by loosening the screw terminals, or by cutting wires to within 2" of sockets.

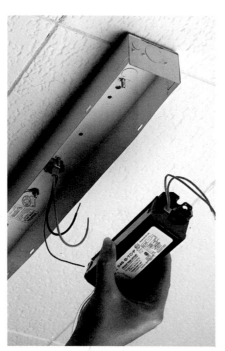

4 Remove the old ballast, using a ratchet wrench or screwdriver. Make sure to support the ballast so it does not fall.

5 Install a new ballast that has the same ratings as the old ballast.

6 Attach the ballast wires to the socket wires, using wire nuts, screw terminal connections, or push-in fittings. Reinstall the coverplate, fluorescent tube, and diffuser. Turn on power to the light fixture at the main service panel.

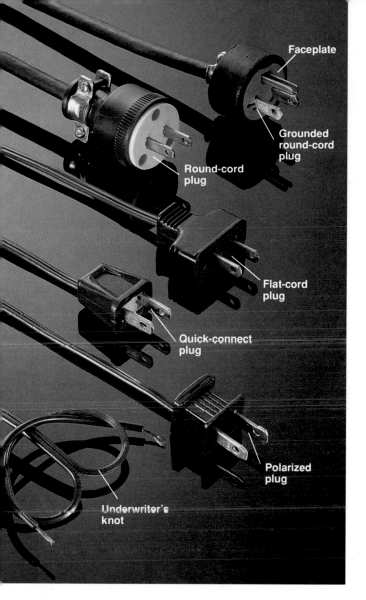

Faceplate

Grounded round-cord plug

Round-cord plug

Flat-cord plug

Quick-connect plug

Polarized plug

Underwriter's knot

Replacing a Plug

Replace an electrical plug whenever you notice bent or loose prongs, a cracked or damaged casing, or a missing insulating faceplate. A damaged plug poses a shock and fire hazard.

Replacement plugs are available in different styles to match common appliance cords. Always choose a replacement that is similar to the original plug. Flat-cord and quick-connect plugs are used with light-duty appliances, like lamps and radios. Round-cord plugs are used with larger appliances, including those that have three-prong grounding plugs.

Some appliances use polarized plugs. A polarized plug has one wide prong and one narrow prong, corresponding to the neutral and hot slots found in a standard receptacle. Polarization (page 12) ensures that the cord wires are aligned correctly with the receptacle slots.

If there is room in the plug body, tie the individual wires in an underwriter's knot to secure the plug to the cord.

Everything You Need:

Tools: combination tool, needlenose pliers, screwdriver.

Materials: replacement plug.

How to Install a Quick-connect Plug

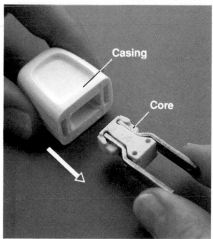

Casing

Core

1 Squeeze the prongs of the new quick-connect plug together slightly and pull the plug core from the casing. Cut the old plug from the flat-cord wire with a combination tool, leaving a clean-cut end.

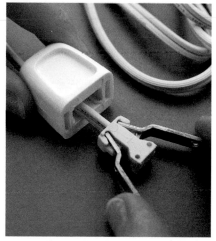

2 Feed unstripped wire through rear of plug casing. Spread prongs, then insert wire into opening in rear of core. Squeeze prongs together; spikes inside core penetrate cord. Slide casing over core until it snaps into place.

Ridged half

Wide prong

When replacing a polarized plug, make sure that the ridged half of the cord lines up with the wider (neutral) prong of the plug.

How to Replace a Round-cord Plug

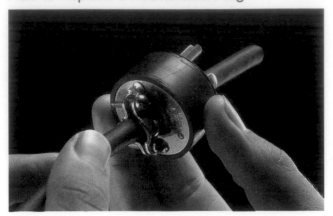

1 Cut off round cord near the old plug, using a combination tool. Remove the insulating faceplate on the new plug, and feed cord through rear of plug. Strip about 3" of outer insulation from the round cord. Strip ¾" insulation from the individual wires.

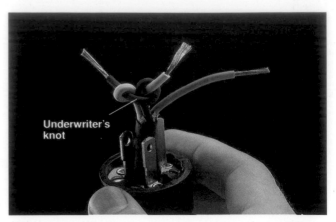

2 Tie an underwriter's knot with the black and white wires. Make sure the knot is located close to the edge of the stripped outer insulation. Pull the cord so that the knot slides into the plug body.

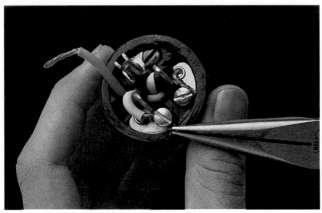

3 Hook end of black wire clockwise around brass screw, and white wire around silver screw. On a three-prong plug, attach third wire to grounding screw. If necessary, excess grounding wire can be cut away.

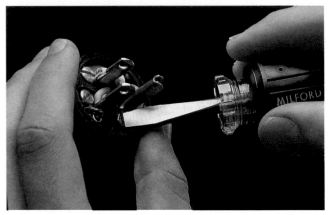

4 Tighten the screws securely, making sure the copper wires do not touch each other. Replace the insulating faceplate.

How to Replace a Flat-cord Plug

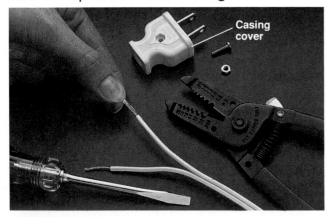

1 Cut old plug from cord using a combination tool. Pull apart the two halves of the flat cord so that about 2" of wire are separated. Strip ¾" insulation from each half. Remove casing cover on new plug.

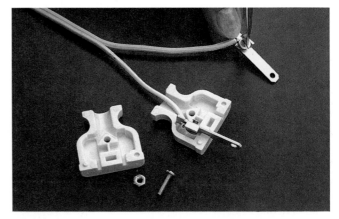

2 Hook ends of wires clockwise around the screw terminals, and tighten the screw terminals securely. Reassemble the plug casing. Some plugs may have an insulating faceplate that must be installed.

Index

Creative Publishing International, Inc.
offers a variety of how-to books.
For information write:
   Creative Publishing Inter-
   national, Inc.
   Subscriber Books
   5900 Green Oak Drive
   Minnetonka, MN 55343